THE BONDING OF BRICKWORK

THE
BONDING OF BRICKWORK

BY

WILLIAM FROST
F.I.S.E., M.R.San.I., M.R.S.T.

Lecturer and Instructor in Brickwork
at the
London County Council
"School of Building",
Brixton

Author of: "The Modern Bricklayer"
"Constructive Sanitary Work" *and*
"Brick Arches, their Setting-out and
Construction"

CAMBRIDGE
AT THE UNIVERSITY PRESS
1933

CAMBRIDGE
UNIVERSITY PRESS

University Printing House, Cambridge CB2 8BS, United Kingdom

Cambridge University Press is part of the University of Cambridge.

It furthers the University's mission by disseminating knowledge in the pursuit of
education, learning and research at the highest international levels of excellence.

www.cambridge.org
Information on this title: www.cambridge.org/9781316603826

© Cambridge University Press 1933

First published 1933
First paperback edition 2015

A catalogue record for this publication is available from the British Library

ISBN 978-1-316-60382-6 Paperback

CONTENTS

THE BONDING OF BRICKWORK

THIS volume deals solely with the bonding of brickwork and has been compiled with the object of providing information for the use of students, craftsmen, foremen and others engaged in the Building Industry, and it is hoped that the professional men who direct the work of the Industry, or are associated with it, will also find the book useful for reference.

The volume contains over five hundred examples of bonding, embracing the principles of setting out bonds for a great variety of conditions and different classes of work.

In the study of bonding it is necessary to obtain considerable practice in arranging the bricks to obtain a desired result. This can be done either by drawing or by actually building alternate courses by patterns or by model bricks.

At the commencement of the illustrations to this volume there is given a set of line drawings, showing, to a scale of 1 inch to 1 foot, the "bed outlines" of the various forms of cut brick which are utilised for the purpose of obtaining the "bond" under practical conditions. These bed outlines are intended to illustrate patterns, which should be re-drawn to a large scale—or even full size—and cut out in stout cardboard, in such numbers as to allow alternate courses of any example to be set out, one upon the other. In a similar way any new idea of bonding may be tried, or any form of bonding demonstrated.

At the same time the student should not neglect drawing, and many examples should be drawn to scale until facility is obtained in arranging the brick outlines on the paper to an accurate scale to ensure a satisfactory bond.

The practice drawings need not have double-line joints as in the book illustrations. The double lines are given for making the outlines perfectly clear; *for practice, the joints should be done in single lines.*

At the end of the book will be found a number of examples for practice in bonding. These consist of the outlines of portions of straight walls, piers, quoins, junctions, irregular and broken plans, etc. In each case the student should attempt to fill in the appropriate setting out of the bricks in both Old English and Flemish bonds.

In practice all kinds of variations in form or in dimensions will occur. The student should observe examples which differ from his previous studies, make a record of their form and dimensions, work out the bond and add the examples to his collection.

BOND

Bond, in relation to brickwork, means the arrangement of the bricks whereby:

(*a*) an adequate distribution of load is obtained through the mass of brickwork;

(*b*) the mass is tied together so that any individual brick is not easily displaced;

(*c*) some uniform and pleasing arrangement of the faces of the bricks appears on the face of the wall.

Bricks vary in size in different parts of the country, but they approximate to a standard of 9″ long, $4\frac{1}{2}$″ wide and 3″ thick, measured when laid in mortar, centre to centre of the joints. Northern and Midland bricks often measure $9\frac{1}{4}$″ × $4\frac{5}{8}$″ × $3\frac{1}{4}$″ centre to centre of the joints when laid. In every case the 9″ × $4\frac{1}{2}$″ dimension is the normal *bed* of the brick—laid horizontal—for ordinary walling, and all units cut for bonding purposes, as shown in the illustration facing page 24, maintain their thickness of 3″ (or other thickness), the line pattern for cutting being the form of the *bed*.

It should be noted that for obtaining accurate bonding —maintaining a uniform overlap of brick upon brick— the length of the brick must be equal to "twice the width plus the thickness of a mortar joint".

KINDS OF BOND

The two main bonds in use for good class work are *Old English bond* (or English) and *Flemish bond*. Their distinctive difference is in the face appearance; English bond shows alternate *courses* of headers (ends) and stretchers (sides) in elevation; Flemish bond shows *alternate bricks in the same course* as header and stretcher, and the header of one course must lie in the centre of the stretcher in each of the courses above and below it.

The difference is illustrated by Figs. 1 and 2.

There are several bonds which have acquired special names amongst builders, but they are all variations upon the Old English and Flemish bonds. These include English garden and Flemish garden wall bonds; in the former, Fig. 3, three to five courses of stretchers are used to one course of stretchers, and in the latter, Fig. 4, one header is succeeded by three (or more) stretchers in the same course.

Stretching bond, Fig. 5, has all bricks laid as stretchers, giving the maximum overlap of $4\frac{1}{2}''$ in the direction of the length. It is suitable for walls $4\frac{1}{2}''$ thick, and for facing thick walls with a special kind of facing brick. When used for the latter purpose, bond between facing and backing is obtained by metal ties embedded in the joints.

Heading bond, Fig. 6, is sometimes used for effect in panelled work and the like, and is also particularly suited to work which is curved on plan, the short length of the header face allowing a reasonable approach to the required curve (if not too sharp) without cutting the bricks.

Brick-on-edge heading bond, Fig. 7, and brick-on-edge stretcher bond, Fig. 8, are appropriate for thick and thin walls respectively. The wall of Fig. 8 is only $3''$ thick (sometimes employed for partitions), while the wall of Fig. 7 may be of any thickness from $9''$ upwards and bonded similarly to other walls, on plan.

Dutch bond, Fig. 9, avoids the use of closers in starting the bond. Three-quarter bats are used in the stretching courses to obtain the position of the first stretcher joint, and a header is inserted after this three-quarter bat *in alternate courses* which breaks the continuity of the perpends (vertical joints) so that they occur only in alternate stretching courses.

English cross bond, Fig. 10, inserts a header after the first stretcher in each stretching course.

Mixed garden-wall bond, Fig. 11, has a true Flemish course mixed with three (or more) courses of stretchers.

Wood-slip bonding, Fig. 12, is done by inserting a length of thin wood slip into a horizontal joint to improve the longitudinal bond, which, except in stretching bond walls, is only an effective $2\frac{1}{4}''$ against sidewise withdrawal. Wood slips are effective and allow of nailed fastenings, but are liable to decay and not much used in modern work. If used at all, they should be well-seasoned dry material, and should be bedded in cement mortar which hardens and dries out speedily.

For modern *bonding* wire mesh is employed.

English Bond

Figs. 13 to 16 show walls $9''$ to $22\frac{1}{2}''$ thick, with square stopped ends—or returns. In the $22\frac{1}{2}''$ wall—Fig. 16—the irregularity of the arrangement of the stopped ends should be observed. This bond is suitable where the lower edges of the plans shown represent the face of the wall—to be seen—and assumes that the back will be plastered or occupy an unimportant position. Alternative methods are available, e.g. Fig. 16 (*a*), but they involve more cutting and are not so efficient for supporting load; they may, however, improve the face appearance.

Flemish Bond

Figs. 17 to 20 show walls $9''$ to $22\frac{1}{2}''$ thick, with square stopped ends—or returns. It will be observed in Fig. 18 —$13\frac{1}{2}''$ wall—that a bonding unit occurs which is $13\frac{1}{2}''$ square with a half bat in the centre.

The length of the wall consists of a number of such units plus "one brick length". A perfect bond is obtained on the face of the wall, but not on the back, an additional header being required to fill up in each course.

The same kind of irregularity occurs in the $22\frac{1}{2}''$ wall—Fig. 20—one face only being perfectly bonded.

ENGLISH BOND WITH FLEMISH FACE

This bond is sometimes called "single Flemish". Figs. 21 to 24 show examples of this bond for walls $13\frac{1}{2}''$ to $36''$ thick.

QUOINS (External angles)

Fig. 25 shows a $4\frac{1}{2}''$ wall in stretching bond. Figs. 26 to 31 show the bonding of quoins in English bond walls from $9''$ to $36''$ thick. The general principle is to allow the stretching face course to run through to the angle and so form a header on the return face, then to butt the whole of the heading course of the return wall against these stretchers. Figs. 32 to 34 show similar quoin bonding applied to walls in Flemish bond from $9''$ to $18''$ thick. Figs. 35 and 36 show quoin bonding for walls $13\frac{1}{2}''$ and $18''$ thick in single Flemish bond. The same principle is adopted for Flemish and single Flemish bonds as for English bond, interpreting the stretching course to mean the course which *commences* with a stretcher.

Cavity walls. Figs. 37 and 38 show the quoin bonding for cavity walls, in the first case with two half-brick walls divided by a $2\frac{1}{4}''$ cavity, and in the second case with the inner wall $9''$ thick. *Note: the cavity may vary in width from $2''$ to $3''$.*

Squint quoins. When the quoins are formed by walls not at right angles they are called *squint quoins.* The angle enclosed may be either acute or obtuse. Acute squint quoins in English bond for walls from 9″ to 36″ thick are shown in Figs. 39 to 43.

The principles are the same as for square quoins but a difficulty occurs because the splayed end of the first stretcher measures more than the normal width of a brick. This brick is therefore reduced by a splay cut (which may be long or short) and a tapered closer is usually inserted, although the exact form may be varied considerably, as shown by comparison of Fig. 39 and the other plans.

The more nearly parallel and the larger the closer, the more satisfactory is the work from the point of view of soundness and practical cutting.

Figs. 44 to 48 show acute squint quoins in Flemish bond for walls 9″ to 36″ thick, and Figs. 49 to 52 give similar quoins for walls in single Flemish bond from $13\frac{1}{2}$″ to 36″ thick. Where Flemish bond is adopted a certain amount of irregular arrangement and cutting will always occur. An examination of the various examples will show however that in every case there is an alternating stepped arrangement in the entry of one wall into the other in the pairs of courses shown.

Figs. 53 and 54 show a more modern idea of arranging the outer angle of a squint quoin. The sharp corner is avoided by allowing an "inset" or "recessed" formation, in which uncut faces of the bricks appear on the recessed faces and the bonding is much sounder and more capable of transmitting loads.

This method is a great improvement over the older form with sharp angles, because the latter required either purposely made "squints" or it caused to be exposed the cut splayed ends of the quoin bricks.

Figs. 55 to 59 show obtuse squint quoins for walls 9″ to 36″ thick in English bond. It should be observed that in all these examples, in order to use an ordinary sized brick for the squint quoin, the header end of the brick is reduced to $2\frac{1}{4}$″ and is followed by a queen closer. The stretcher face is $6\frac{3}{4}$″ long and thus a perfect face bond is established. Fig. 56 (a) shows how the square angle of the quoin brick may be retained to form a dog-tooth quoin by allowing the sharp angle to overhang.

Figs. 60 to 64 show similar obtuse square quoins in Flemish bond for walls from 9″ to 36″ thick and Figs. 65 to 68 show single Flemish obtuse quoins for walls from $13\frac{1}{2}$″ to 36″ thick. The same general principles apply to these cases as for the English bond quoins and earlier examples of squint bonding.

Recessed jambs with $4\frac{1}{2}$″ reveals and $2\frac{1}{4}$″ deep recesses are shown in English bond in Figs. 69 to 73 for walls 9″ to 36″ thick. The main principle is to employ a bevelled half bat and king closer to commence the reveal in the heading course, and then to employ bevelled closers and bats as required to make up the bond in the recesses.

Figs. 74 to 78 show the same range of reveals in English bond with the recesses $4\frac{1}{2}$″ deep, as required for vertical sliding sash boxes. In this case a half bat is used to start the heading course and bevelled closers immediately follow.

Figs. 79 to 81 show English bond with reveals 9″ wide and recesses $4\frac{1}{2}$″ deep. In this case the cutting is entirely behind the face bricks and introduces bevelled closers, bevelled bats and king closers, all being obviously simple in their use and suitability.

A similar complete series of examples is given in Figs. 82 to 84 for $2\frac{1}{4}$″ recesses and $4\frac{1}{2}$″ reveals in Flemish bond, and in Figs. 85 to 88 with $2\frac{1}{4}$″ recesses and 9″ reveals. Figs. 89 to 92 show $4\frac{1}{2}$″ recesses and 9″ reveals, also in Flemish bond.

Fig. 93 shows how to form a $13\frac{1}{2}$″ reveal with a $2\frac{1}{4}$″ recess in English bond. This is sometimes required in deeply recessed doorways.

ATTACHED PLAIN PIERS

Figs. 94 to 98 show the bonding of attached plain piers with $2\frac{1}{4}$″ projection to 9″ walls in English bond, for piers varying from 9″ to 36″ wide.

Observe that the heading course of the pier enters the stretching course of the wall in each case, and that bevelled cutting only occurs in the stretching course of the pier.

Figs. 99 to 101 show similar piers, $2\frac{1}{4}$″ projection and $13\frac{1}{2}$″ to $22\frac{1}{2}$″ wide, attached to walls in English bond, $13\frac{1}{2}$″ thick; and a similar series in 18″ walls is shown by Figs. 102 to 104.

Figs. 105 to 114 show similar ranges of attached piers $4\frac{1}{2}$″ projection in English bond for $13\frac{1}{2}$″ and 18″ walls.

Flemish bond. Attached piers in Flemish bond walls are illustrated for various widths and $2\frac{1}{4}''$ projection and in walls from 9″ to 18″ thick in Figs. 115 to 124, and piers in single Flemish bond walls $13\frac{1}{2}''$ thick are similarly illustrated in Figs. 125 to 127. Piers having a projection of $4\frac{1}{2}''$ are shown for Flemish bond and single Flemish bond, over a similar range to the above, in Figs. 128 to 143.

In all these piers a similar principle is followed, as described above, except that where the bonding is to Flemish bond walls, bevel bats and king bats and closers are employed: reference to particular examples will make this clear. It should be noted that all the above examples assume that the piers occur in suitable positions to facilitate bonding. In practice many variations may occur when, through faulty work or faulty selection of positions or spacing dimensions, the pier does not come in line with appropriate joints; much unnecessary cutting may then result. The correct thing is to ensure that the bonding surrounding the pier is correctly set out, to work outwards from the pier in each direction and to allow any necessary broken bond to occur in the centres of the bays between the piers. Where work is judiciously planned, with well-selected dimensions, and the work is executed by a good craftsman, it should always be possible to adjust the bonding in a satisfactory way which at least approaches to the ideal bond shown in the examples described above.

Attached piers with chamfered angles. These need no special description as they are bonded on principles already understood by the student. Chamfer bricks with

$2\frac{1}{4}''$ splays are used; these are of standard pattern. The least projection of the pier is $4\frac{1}{2}''$ and Figs. 144 and 145 show examples of such piers $13\frac{1}{2}'' \times 4\frac{1}{2}''$ projection, and $18'' \times 9''$ projection, which are suitable for load carrying under floor beams, roof trusses, etc.

Double Attached Piers

A range of examples of double piers—one on each face of the wall—is shown in Figs. 146 to 152. In some of these examples where the projection is $4\frac{1}{2}''$, it is more convenient to place the stretching course of the pier in the stretching course of the wall. Such cases call for some initiative on the part of the student and craftsman, and there may be quite satisfactory alternative ways of solving the problem. Good architectural appearance is an important consideration in all this work where the face of the wall is to be left exposed. Jointing, its disposition and balance, is therefore to be considered.

Figs. 153 to 155 are a few special cases of attached piers.

Junctions

Figs. 156 to 160 show right-angled (or square) junctions in English bond for walls from $9''$ to $36''$ thick.

The principle in every case is to allow the heading course of the abutting wall to enter $2\frac{1}{4}''$ into the stretcher face of continuous wall and to make up the remaining $2\frac{1}{4}''$ by queen closers. The effective bond is therefore $2\frac{1}{4}''$, which is really the maximum amount for this bond, and occurs at every alternate course.

Figs. 161 to 165 show a similar series in Flemish bond. The principle is the same as that adopted for English bond junctions, with slight variations and adjustments which are required by the alternating header and stretcher in the same course.

Figs. 166 to 169 illustrate suitable junctions between walls in which *one face of each* is in Flemish bond.

Splay junctions. The principle of joining two walls which meet on the splay (any angle other than a right angle) is to adapt the previous method of allowing the heading course of the abutting wall to enter the stretching face of the continuous wall by at least $2\frac{1}{4}''$, and to butt the stretching course (by splayed cutting) against the continuous wall. Awkward cutting cannot be avoided in these cases, and the endeavour of the craftsman should be (*a*) to ensure an adequate tie between the walls, and (*b*) to use as few small pieces of cut brick as possible. Note that in the thicker walls the entry of the heading course of the abutting wall is often carried to the back of the face stretchers.

Figs. 170 to 175 show examples of splayed bonding applied to walls in English bond.

Cross junctions. Where two walls (of the same or of different thicknesses) cross each other at right angles, alternate courses run through. Figs. 176 to 181 show acceptable arrangements for the bonding of such walls. The student should practise similar bonding for walls of different thicknesses and walls on the splay, and should note that considerable variation in practical conditions may occur according to the exact position at which the crossing takes place.

Dimensions should be selected where possible, therefore, to allow perfect bond to be arranged.

While these examples of cross junctions include both English and Flemish bonds, it should be observed that such junctions in Flemish bond are not frequently required.

Junctions of curved and straight walls (radius junctions). Figs. 182 to 185 show junctions between straight and curved walls in English and Flemish bonds and incidentally show the arrangement of the bonding in curved walls, using uncut bricks to obtain the approximate plan curve. It should be noted that quick curves are most appropriately carried out in heading bond if uncut bricks are to be used, as the small width of the header allows a closer approximation to the curve. The back of such a wall cannot however be exposed because the joints are very wide. If both faces *must* be exposed, then the bricks must be axed to a wedge shape or special (purpose made) bricks be employed.

"BREAKS" OR "RETURNS"

Where a break in a wall occurs at a projection or recess, two angles are formed, viz. an external and an internal quoin. If these are in close association—with only a short wall between—the pair of quoins and their short connecting wall are often referred to as a Z junction.

Figs. 186 to 190 show such junctions in walls of uniform thickness in English bond. The break or return length should be a definite number of half bricks.

A further series, where the break or return is equal to the thickness of the wall, is given in Figs. 191 to 194.

Detached or Isolated Piers

The bonding of detached piers is usually very simple and direct when the bonding of walls has been previously understood.

Figs. 195 to 200 show piers from 9″ to 36″ square in English bond. The alternate courses are identical in form if correctly bonded; they are merely reversed in direction. Note that $13\frac{1}{2}$″ and $22\frac{1}{2}$″ piers cannot be arranged in *perfect* English bond because of the odd half brick. The $13\frac{1}{2}$″ pier is shown with alternative methods of bonding in Figs. 196 and 197. Fig. 196 gives the nearest approach to English bond but all the bricks are cut down to $\frac{3}{4}$ bats. Fig. 197 involves much less cutting and is satisfactory.

Flemish bonding of square piers is shown in Figs. 201 to 204. The $13\frac{1}{2}$″ square pier, Fig. 201, is remarkably strong, although the centre $\frac{1}{2}$ bat is merely a filling, because of the $4\frac{1}{2}$″ overlap of the facing bricks. The 36″ square pier, Fig. 204, is not symmetrical in the arrangement of each course but gives a strong and acceptable bond.

Chimney-stack Bonds

The bonding of ordinary chimney stacks containing series of flues from domestic fireplaces is shown in Figs. 205 to 210. The first two examples show 9″ square and $13\frac{1}{2}$″ × 9″ flues respectively in English bond, using 9″ external walls. Fig. 207 shows $13\frac{1}{2}$″ × 9″ flues with a $4\frac{1}{2}$″ external wall built in Flemish bond, which requires $\frac{1}{2}$ bats or snap headers between the stretchers. Fig. 208 shows the same flues with 9″ external walls in Flemish bond. The bond is necessarily irregular, two headers occurring in succession on the long face.

Domestic stacks are most usually built with walls $4\frac{1}{2}''$ thick, and "stretching" or "chimney" bond is used. The arrangement of the bond is then remarkably simple. Figs. 209 and 210 show this bond with different numbers and grouping of flues. In Fig. 209 there are two 9" square and two $13\frac{1}{2}'' \times 9''$ flues, while in Fig. 210 there are four 9" square and two $13\frac{1}{2}'' \times 9''$ flues. The arrangement, with its broken faces, makes for good architectural effect.

Fireplace Openings

Figs. 211 to 213 show the bonding of the jambs to fireplace openings, the jambs varying in width and the fireplace openings in depth.

Fig. 211 shows a 9" deep recess and 9" wide jambs, the opening being 3' 9" wide and formed in a 9" wall which maintains the full thickness at the back of the opening.

Fig. 212 shows a 3' opening formed in a $13\frac{1}{2}''$ wall, the jambs being 18" wide (to receive a flue), the recess $13\frac{1}{2}''$ deep and the "back" 9" thick. A 9" square flue is shown with 9" of brickwork behind it—a good provision for an external wall.

Fig. 213 shows back-to-back fireplace openings, 3' wide, in a $13\frac{1}{2}''$ party wall, with 9" solid work between the fireplace recesses; 9" square flues are shown but $13\frac{1}{2}'' \times 9''$ flues can be inserted if required.

RIGHT-ANGLED QUOINS WITH STOPPED ENDS AND REVEALS

Figs. 214 to 217 show two short walls forming a quoin, in each case, with a stopped end on one arm and a recessed jamb forming a reveal on the other.

SPECIAL FORMS OF STEPPED REVEALS

Jambs are sometimes formed in a number of recessed steps or reveals in order to widen a window opening on the inside. The same formation is also used for certain types of doorway, but in reverse order, on the outside.

Figs. 218 to 225 show stepped reveals for various walls in English and Flemish bonds.

PROJECTING QUOINS

Solidity of appearance and extra weight is often given to quoins by giving them $2\frac{1}{4}''$ or more projection and thereby forming an attached pier, or double pilaster, at the angle.

Various cases, with their appropriate bonding, are shown in Figs. 226 to 230.

The term "block quoins" is often used for the above.

MISCELLANEOUS EXAMPLES OF BONDING

A large number of examples of a varied character are given in Figs. 231 to 253. They include unusual or irregular lay-outs, embodying piers, pilasters, insets or panels, returns, breaks, splayed junctions, etc. The student may easily obtain their dimensions by noting the numbers and kinds of brick units which compose their parts and should set the outlines out to scale and attempt their bonding as an exercise.

At this stage it is unwise merely to copy the examples given. More lasting information will be retained by first attempting the problem and then checking the result by the solution given.

Note also that there may be other and quite satisfactory methods of arranging the bond for any one of these unusual examples.

BROKEN BOND

In carrying out work where openings occur between piers of brickwork, it sometimes happens, either through force of circumstances or by the selection of dimensions that will not allow of correct bonding, that some form of broken bond may have to be used.

Fig. 254 shows the alternate course of a 9″ wall in English bond *above* or *below* a 2′ 5¼″ opening, and assumes piers to the left and right of this opening which are respectively 2′ 3″ and 2′ 7½″ wide. Headers and ¾ bats have to be inserted in the stretching course and ¾ bats in the heading course, thus forming a broken bond.

Fig. 255 shows a case of irregular dimensions in the opening sizes—1′ 8¼″ and 1′ 1½″—with piers 1′ 6″ and 1′ 10½″ wide in a 13½″ wall in English bond. The brickwork above or below the openings requires to be broken in order to give *correct bond at the piers*, as shown by the plans of the alternate courses.

Fig. 256 shows broken courses in a 13½″ wall in Flemish bond, and Figs. 257 and 258 similar cases for a 9″ wall in Flemish bond. The need for the broken bond is assumed to arise from the fact that the selected *length* of wall does not allow of perfect face bond.

Figs. 259 and 260 illustrate broken bond which is necessary for a pier 3′ 6¾″ on face, with one square stop and one 4½″ recessed reveal. The walls are 13½″ thick and the example is given in (*a*) English, and (*b*) Flemish bond.

Cavity Walls

Recessed (or plain) jambs in cavity walls for good work are made solid at the sides of the opening. Figs. 261 to 264 show pairs of alternate courses indicating the bond for 4½″ reveals and 2¼″ recesses, and for 4½″ reveals and 4½″ recesses. The cavities are all 2¼″ wide.

Figs. 261 and 263 are the normal type of cavity wall with inner and outer portions 4½″ thick. Figs. 262 and 264 show the inner portion 9″ thick.

In order to prevent the transmission of moisture to the inner wall from the outer, the bricks closing the gap should be non-absorbent or the joint abutting against the outer wall should be tarred and/or bedded in non-absorbent material.

Square-angled Bay Windows

Figs. 265 and 266 show the bonding of square-angled bay windows in English and Flemish bonds. The walls are 9″ thick and the width of the bay should be such as to allow accurate and unbroken bonding.

In English bond a multiple of the length of a brick will always be satisfactory, and by the insertion of a central header in the stretching course the variation may even be by half-brick lengths.

In Flemish bond the outside length of the face of the bay must be $22\frac{1}{2}''$+ some multiple of $13\frac{1}{2}''$, e.g. $36''$, $49\frac{1}{2}''$, $63''$, $76\frac{1}{2}''$, and so on, $63''$ ($5'\ 3''$) being about the least dimension which is practicable. *Note: the same rule applies to piers and isolated masses of walling in Flemish bond.* If any other dimensions be employed broken bond must be adopted.

Observe that the two ends of the bay project $18''$ only, hence these portions are built in English bond as the only sound and workable arrangement.

FOOTINGS

The footings of a wall or pier are usually built—as far as possible—in headers. At quoins, and where the courses have an odd number of half bricks in thickness, stretchers have to be introduced. In all cases, however, as many headers as possible are used in order to distribute the load efficiently; half bricks on the outside of a course tend to tilt, as they are only held by the $2\frac{1}{4}''$ which enters the wall.

Fig. 267 shows the two footing courses at the quoin of a $9''$ wall. Bevelled closers are introduced to obtain perfect bond, but when built in cement queen closers are often inserted and in important work the quoin is strengthened by a few pieces of wire mesh embedded in the joints diagonally across or folded round the angle in plan. In long lengths of a "course", such as No. 2, where stretchers must be used, they may be placed with advantage on alternate sides throughout the course.

Fig. 268 shows the three courses of footings for a $13\frac{1}{2}''$ wall, and Fig. 269 the four courses for an $18''$ wall.

Footings to piers are arranged on the same principle but without closers; these are unnecessary. It will be found on examination that some few joints of the *neat work* may coincide with the vertical joints of the top course of footings. (The same thing occurs in quoins.) This is unavoidable and does not cause material weakness so long as the longitudinal bonding is good. (It is wise, in continuous walls in English bond, to place the stretching course first upon the footings.)

Fig. 270 shows the three footing courses (numbered 1, 2 and 3 upwards) and the neat work for a $13\frac{1}{2}''$ square pier, and Fig. 271 shows the four footing courses for an 18″ square pier.

Fig. 272 is a plan of the neat work of the above-named pier and shows the four projecting courses completely assembled.

It should be noted that the base of a pier, being spread in *two* directions, need not be broadened out at the base until its width is twice that of the pier itself, except when the soil is so soft as to demand special treatment. If spread to double the width of the base of the pier—in both directions—its area is *four* times that of the pier. In the case of a continuous wall, where the spread is in *one* direction only, the area of the base is only *twice* that of the wall. Piers are, however, used for isolated loads, and their necessary "spread bases" should usually be determined by calculation for the existing conditions.

Fig. 273 shows the footing courses at the obtuse-angled quoin of an 18″ wall, and Fig. 274 similar footing courses for an acute-angled quoin.

As an exercise the reader should draw the footing

courses separated, showing full detail of the cutting and arrangement of the bricks.

Special cases sometimes arise where unusual arrangements of the bricks must be adopted. This need may arise from a desire to improve the bond usually employed, or from the particular form of the example.

Fig. 275 is a type of bond which has been used to increase the longitudinal bond in a thick wall: examination of the usual arrangement (Fig. 169) will show that the centre portion of the stretching course is filled with headers and the lengthwise tie is therefore only $2\frac{1}{4}''$ in the body of the wall. The arrangement shown in Fig. 275 introduces a complete course of stretchers. This may be inserted in *alternate courses of stretchers*, or only at intervals of 6 to 8 courses, being substituted for the usual stretching course. This type of bond does definitely improve the longitudinal bond, but is expensive because of the large number of closers required. In modern work, reinforcement is the more economical method. This is done by inserting suitable wire mesh in the joints.

Fig. 276 shows the special arrangement of the bond for a $9''$ wall with a projecting plinth, which is broken around an attached pier. Fig. 276 (a) is the elevation, Fig. 276 (b) the plans of the alternate courses of the wall and pier above the plinth, Fig. 276 (c) the plans of the alternate courses *in* the plinth, and Fig. 276 (d) the plan of the top course of the plinth. Note that the plinth walling is $9'' + 2\frac{1}{4}'' = 11\frac{1}{4}''$, hence $\frac{3}{4}$ bats have to be employed to back up the stretchers. An alternative would be to build the plinth as a cavity wall with two $4\frac{1}{2}''$ sections and a $2\frac{1}{4}''$ cavity. Stretching bond would be employed in this case.

Fig. 277 is an arrangement of bricks which might be considered as an alternative to Fig. 275. It is known as diagonal or raking bond. The object is to improve the longitudinal tie. It is wasteful and therefore expensive and has not all the advantages which are sometimes claimed for it. Its chief use in the past was as an occasional insertion in the stretching courses of thick walls and in footings where double-course steps were employed.

Fig. 278 shows a bond called *rat-trap* bond, and suitable for employing bricks-on-edge, thus exhibiting courses $4\frac{1}{2}''$ deep instead of the usual $3''$. This alters the "scale" of the building. The bond gives a greater overlap ($3''$ in lieu of $2\frac{1}{4}''$), there are no closers, and hence the quoin is stronger, if the bricks are good and square. Bricks without frogs are necessary for the face work and wire cuts are therefore suitable.

PATTERN BONDING

True pattern bonding consists of the formation of patterns by selecting bricks having a different colour from the rest of the work and inserting them either in the ordinary bond, or in some special grouping of headers and stretchers, to show a pattern or design on the face.

Two examples are given in Figs. 279 and 280. In the first example the pattern is brought out entirely by headers and therefore requires the insertion of header bricks in the stretching course. The second pattern does not interfere quite so much with the regular bond, but nevertheless is specially arranged for the purpose in view.

BRICKWORK PANELS

Panels of brickwork are often set within straight pieces of plain walling. Figs. 281 to 283 give examples of such panels.

Fig. 281 shows a diaper bond panel: the bricks show their stretcher face but are arranged in groups of three bricks to form squares of 9″ side. This work is effective in panels of moderate size. Fig. 282 shows a panel arranged entirely with headers, the upper and lower courses being on "edge", and therefore $4\frac{1}{2}$″ deep.

The above panels are suitable for walls 9″ or more in thickness.

Fig. 283 shows a panel formed of herring-bone work. All the bricks show as stretchers on the face, but placed obliquely. Such panels are suitable for filling spaces between wood framing, the thickness of the facing being $4\frac{1}{2}$″.

EXAMPLES FOR PRACTICE

The dimensioned figures, Nos. 1 to 42, are prepared in order to enable the student to obtain adequate practice in bonding after studying the methods of bonding which are given in full detail. He should in addition collect and work his own examples.

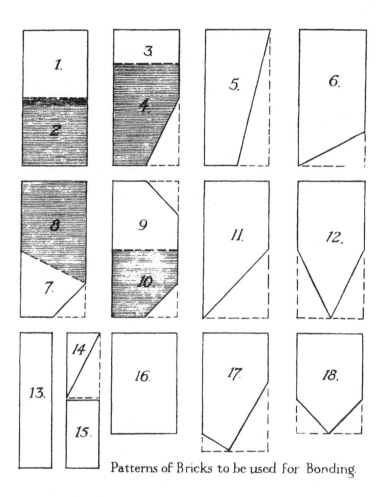

Patterns of Bricks to be used for Bonding.

EXAMPLES OF FACE BONDS
Figs. 1 to 12

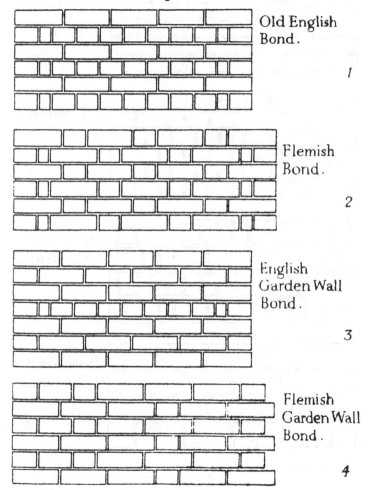

Old English
Bond.

1

Flemish
Bond.

2

English
Garden Wall
Bond.

3

Flemish
Garden Wall
Bond.

4

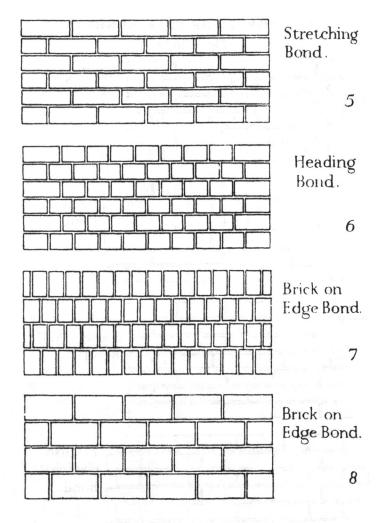

Stretching Bond.

5

Heading Bond.

6

Brick on Edge Bond.

7

Brick on Edge Bond.

8

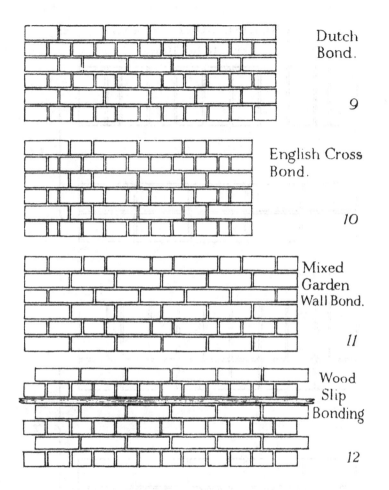

Dutch
Bond.

9

English Cross
Bond.

10

Mixed
Garden
Wall Bond.

11

Wood
Slip
Bonding

12

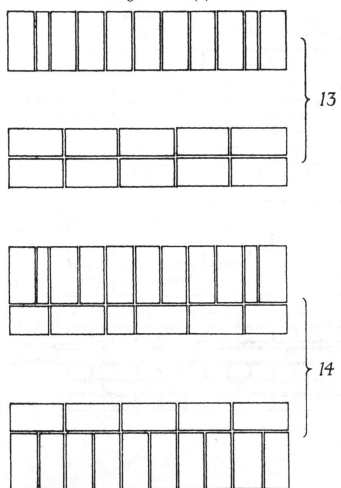

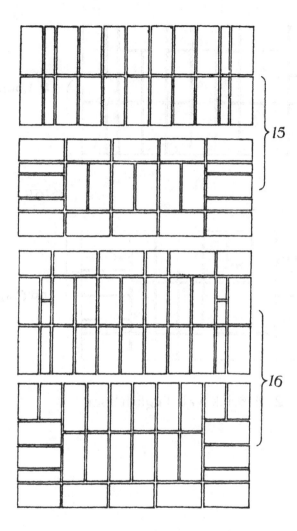

15

16

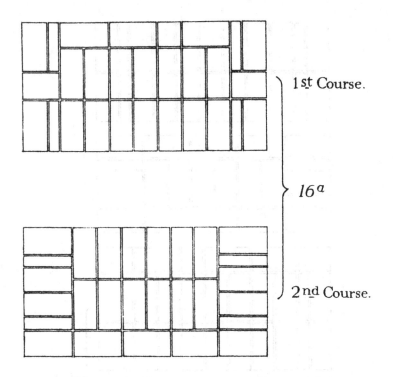

1st Course.

16^a

2nd Course.

22½ Brick wall English Bond

FLEMISH BOND. PLANS

Figs. 17 to 20

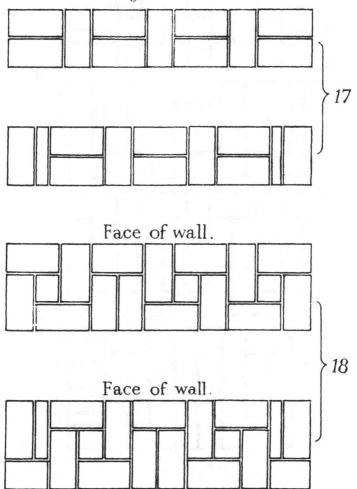

Face of wall.

Face of wall.

17

18

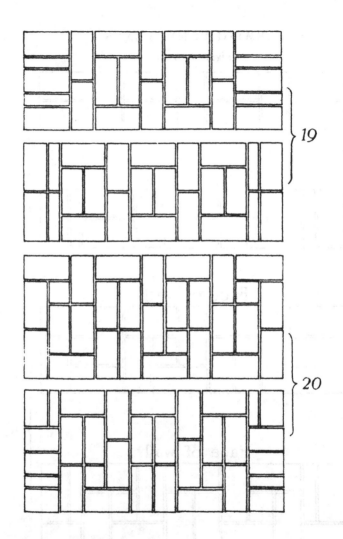

19

20

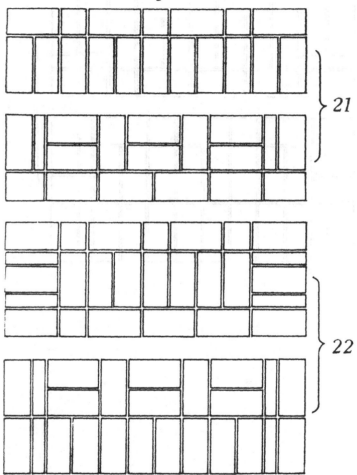

21

22

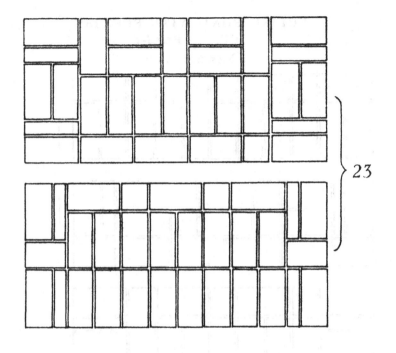

23

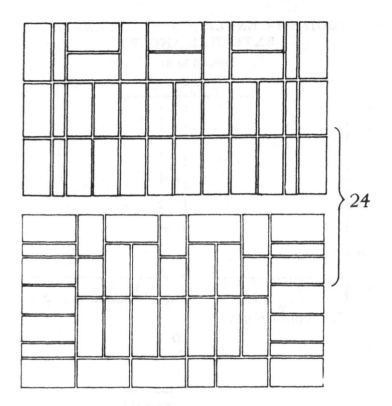

24

QUOINS IN ENGLISH BOND. PLANS OF
EXTERNAL ANGLES

Figs. 25 to 31

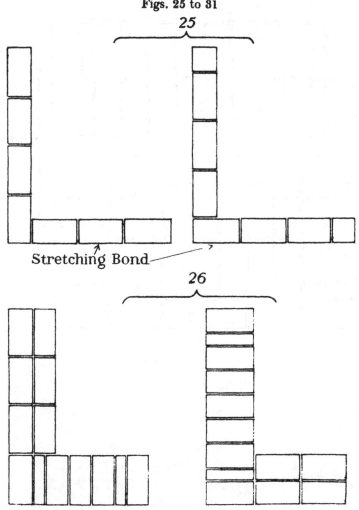

25

Stretching Bond

26

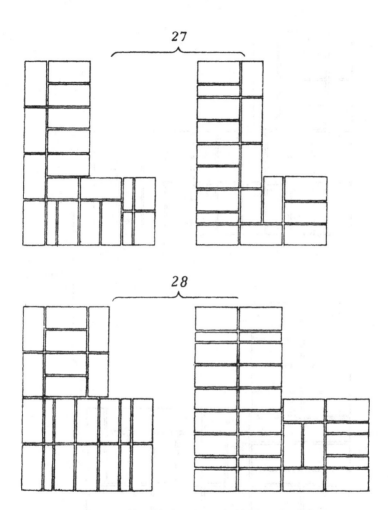

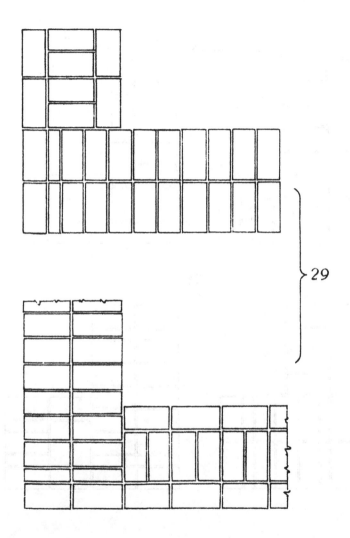

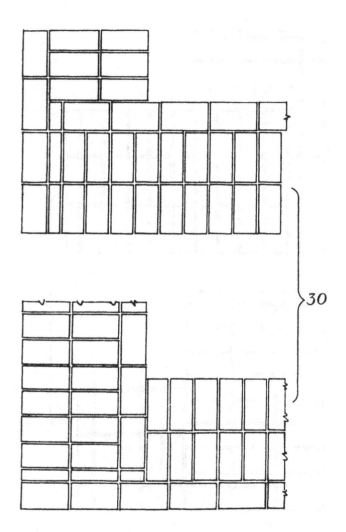

30

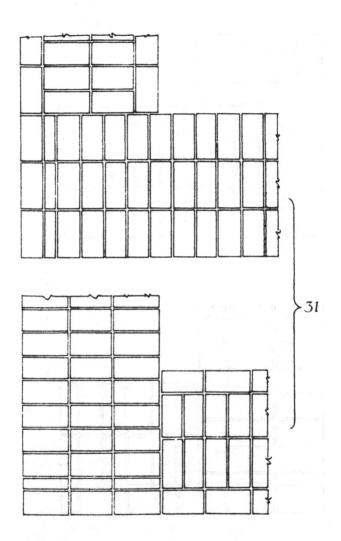

31

QUOINS IN FLEMISH BOND. PLANS OF EXTERNAL ANGLES

Figs. 32 to 34

32

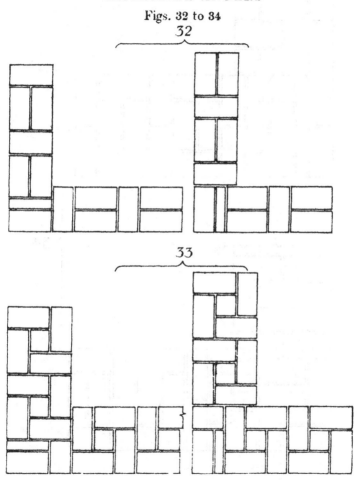

33

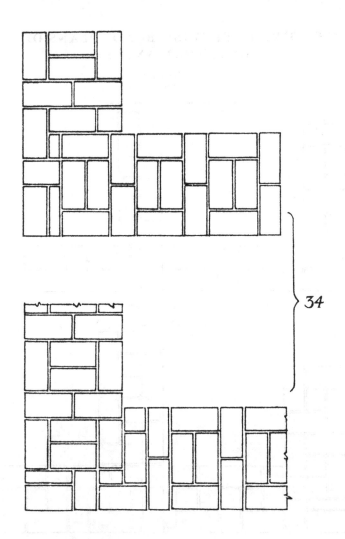

34

QUOINS IN SINGLE FLEMISH BOND.
PLANS OF EXTERNAL ANGLES

Figs. 35 and 36

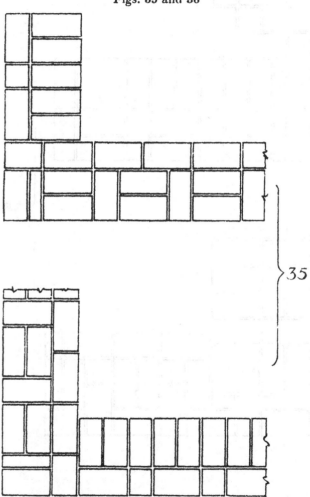

35

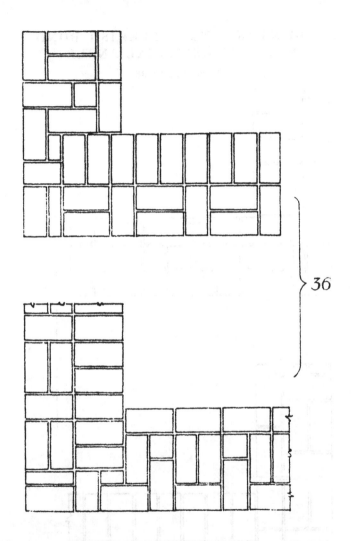

36

QUOIN BONDS FOR CAVITY WALLS.
PLANS OF EXTERNAL ANGLES

Figs. 37 and 38

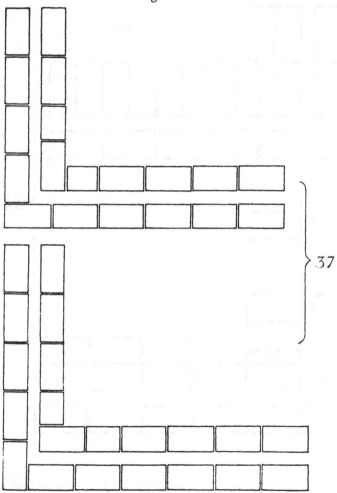

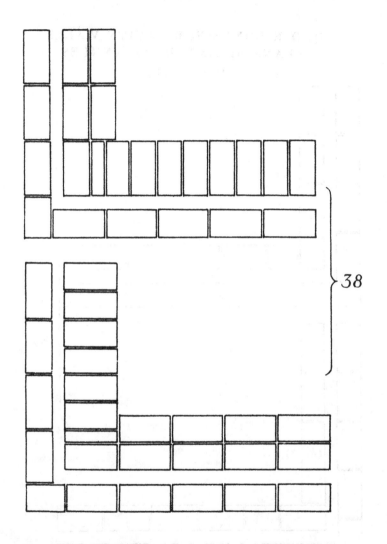

38

SQUINT QUOINS: ENGLISH BOND.
PLANS OF ACUTE ANGLES
Figs. 39 to 43

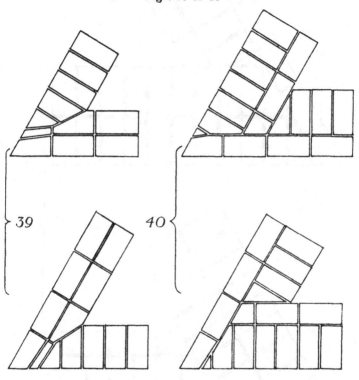

39   40

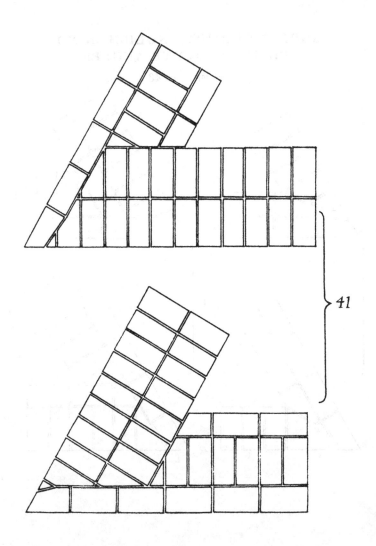

41

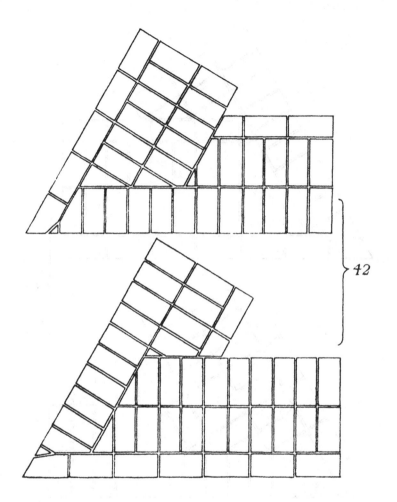

42

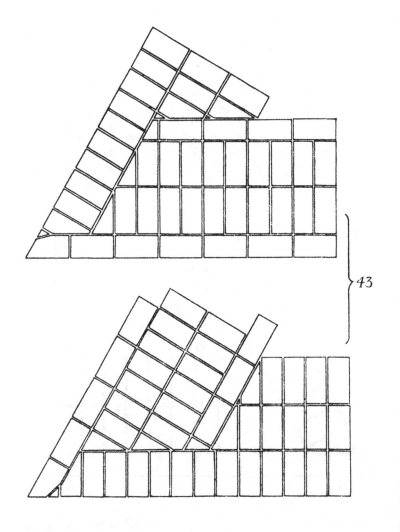

43

SQUINT QUOINS: FLEMISH BOND.
PLANS OF ACUTE ANGLES
Figs. 44 to 48

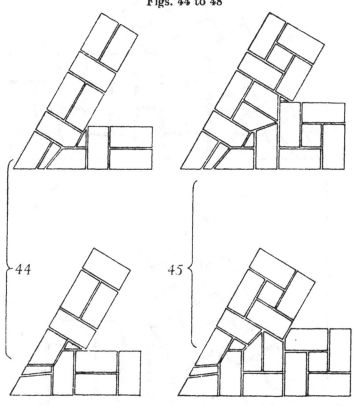

44

45

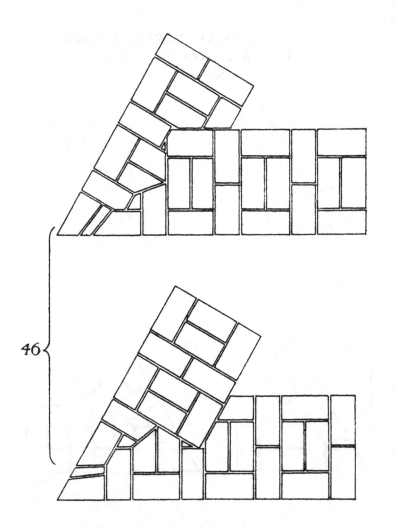

46

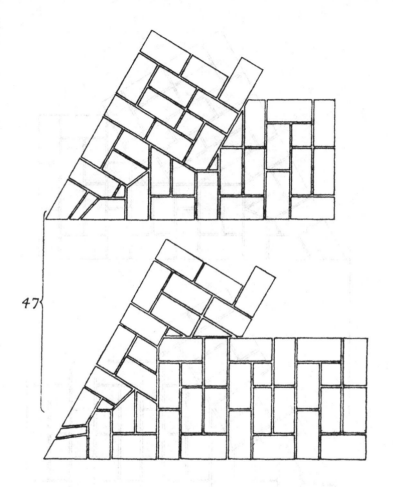

47

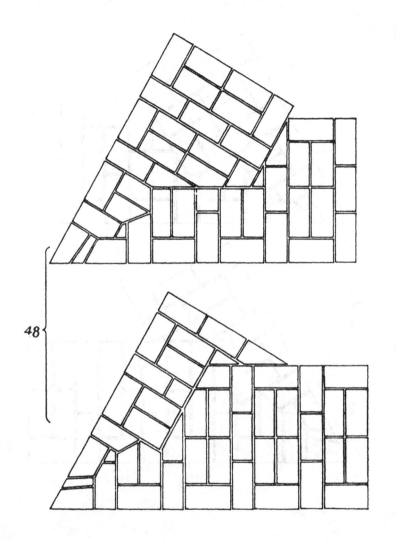

48

Figs. 49 to 52

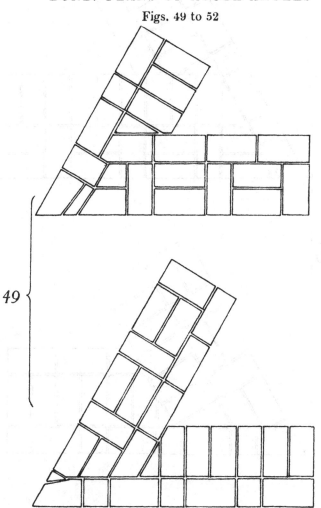

49

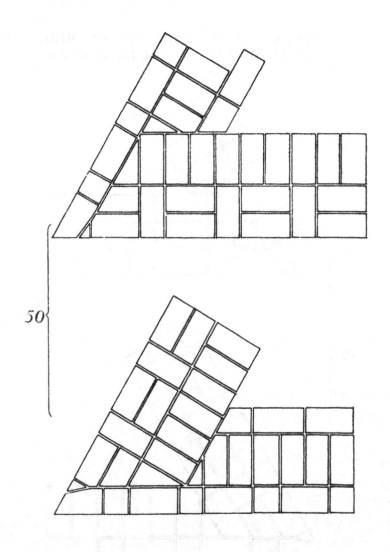

50

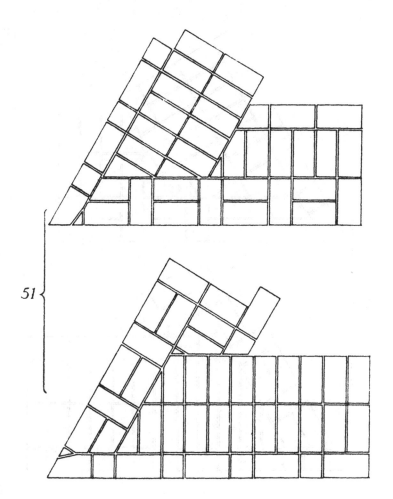

51

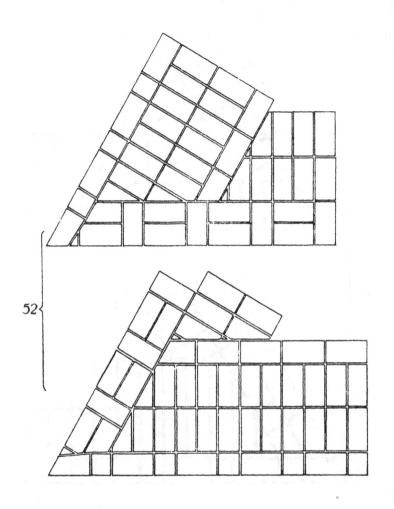

52

SQUINT QUOINS: RECESSED CORNER.
PLANS OF ACUTE ANGLES

Figs. 53 and 54

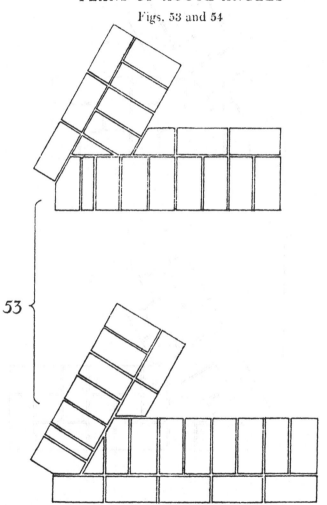

53

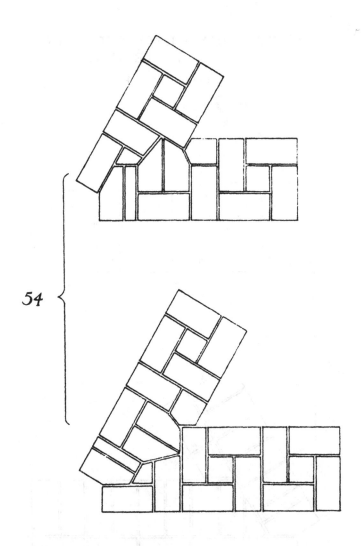

54

SQUINT QUOINS: ENGLISH BOND.
PLANS OF OBTUSE ANGLES

Figs. 55 to 59

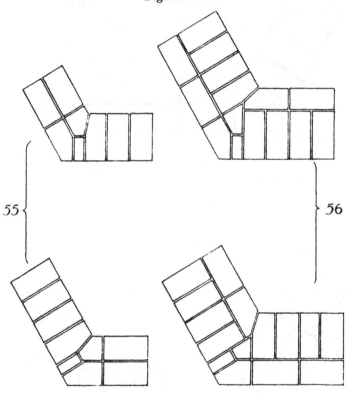

55

56

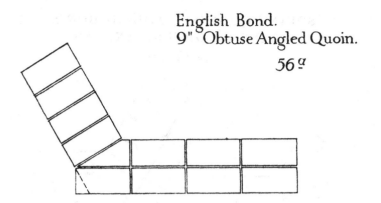

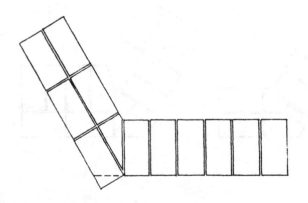

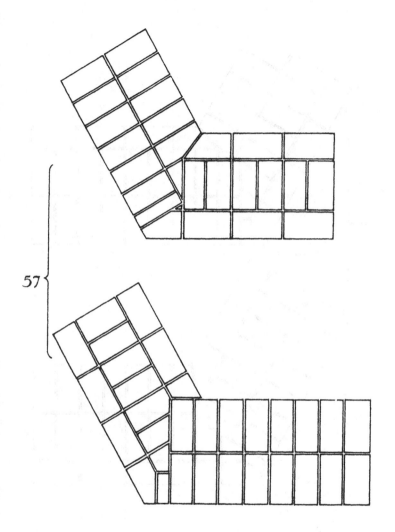

57

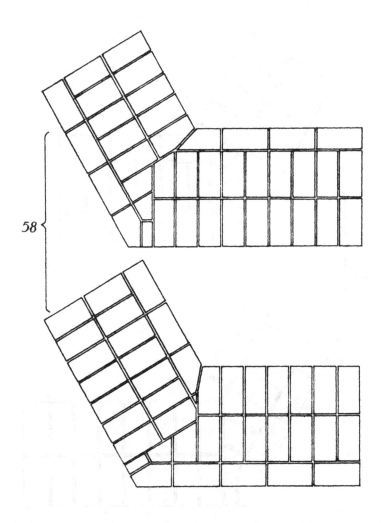

58

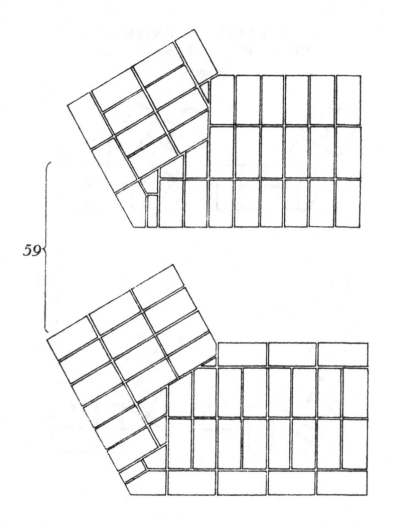

59

SQUINT QUOINS: FLEMISH BOND.
PLANS OF OBTUSE ANGLES

Figs. 60 to 64

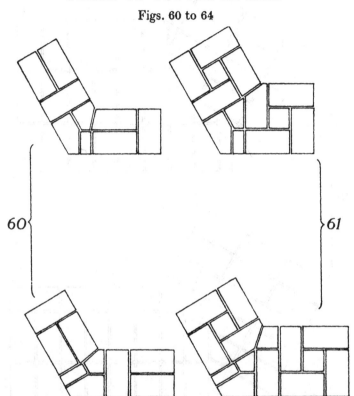

60   61

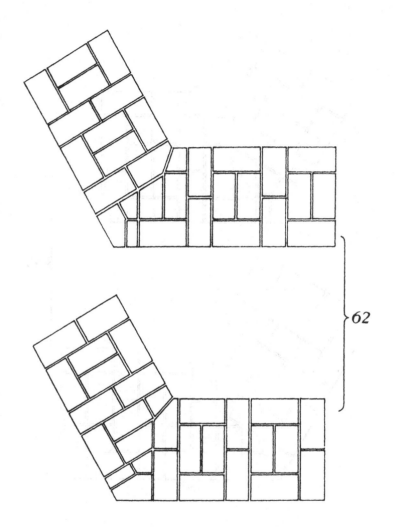

62

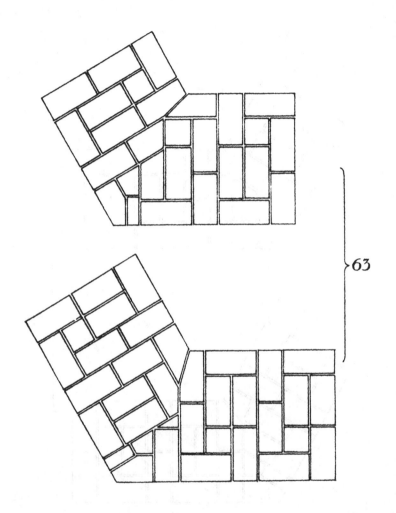

63

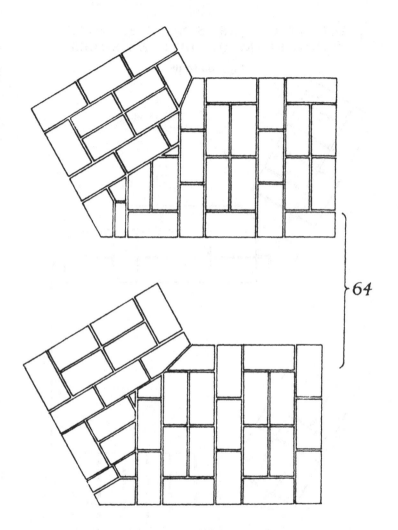

64

SQUINT QUOINS: SINGLE FLEMISH
BOND. PLANS OF OBTUSE ANGLES

Figs. 65 to 68

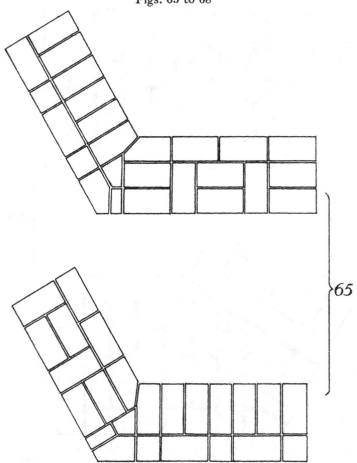

65

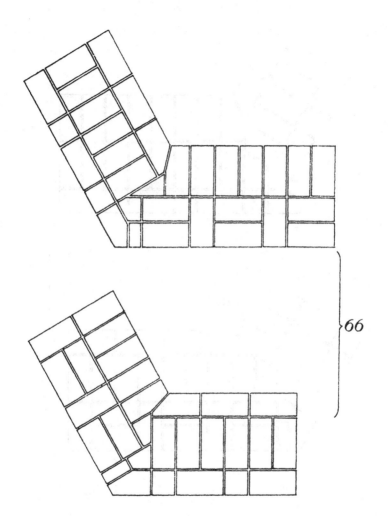

66

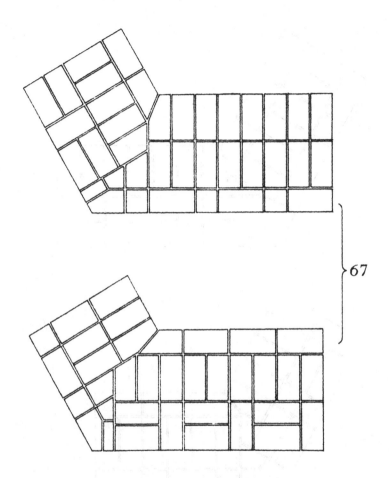

67

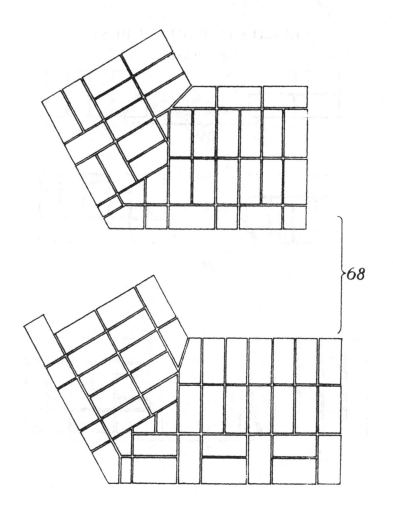

68

REVEALS IN ENGLISH BOND

Figs. 69 to 81

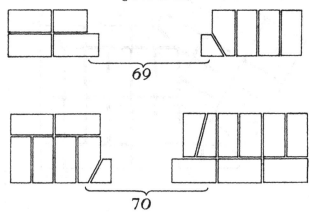

69

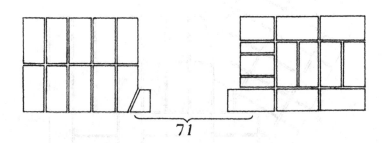

70

71

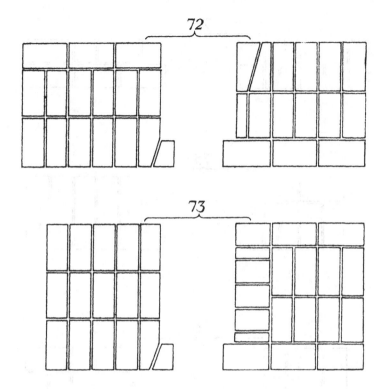

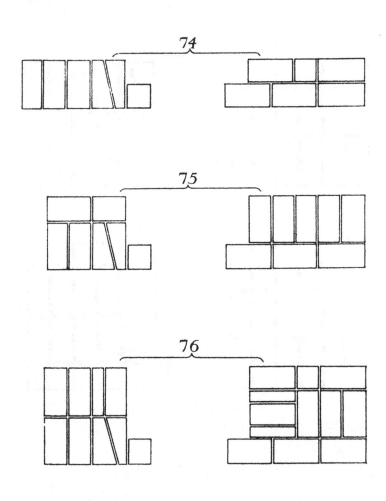

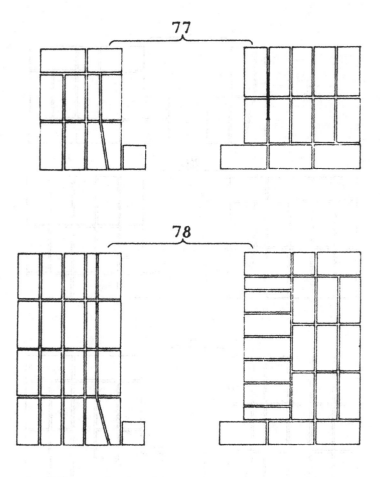

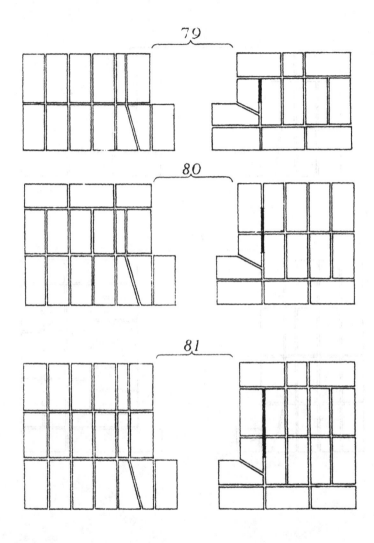

REVEALS IN FLEMISH BOND

Figs. 82 to 93

82

83

84

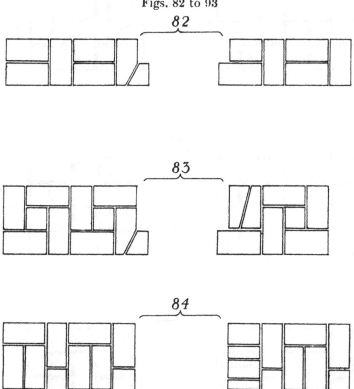

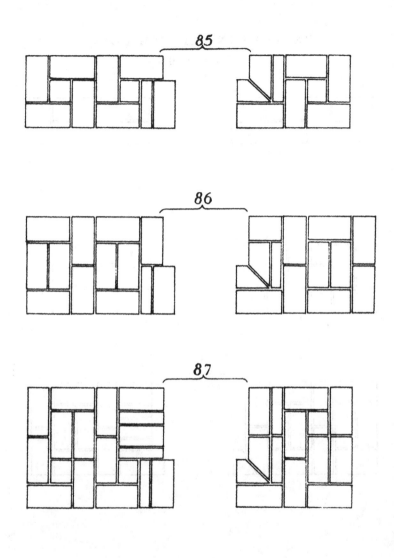

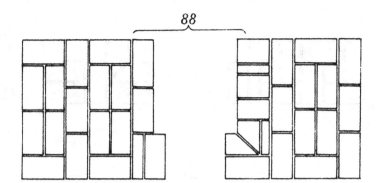

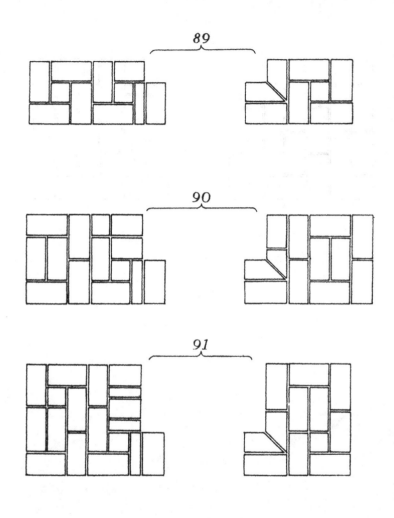

89

90

91

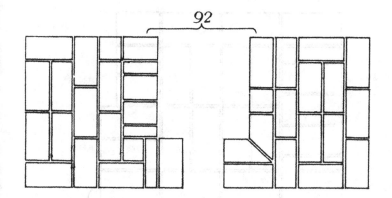

92

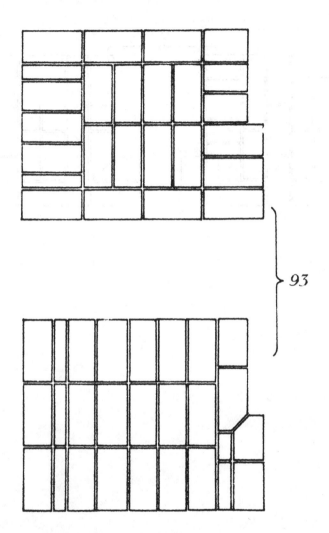

93

ATTACHED PLAIN PIERS IN ENGLISH BOND

Figs. 94 to 114

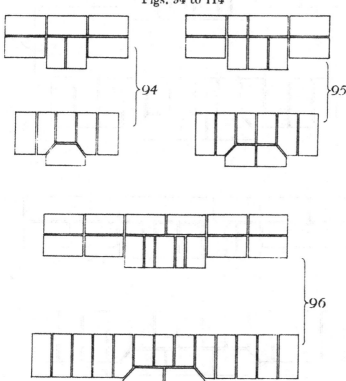

94

95

96

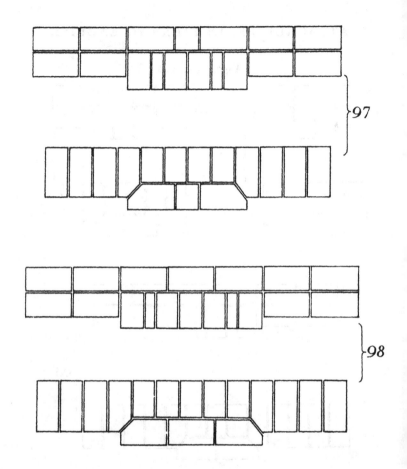

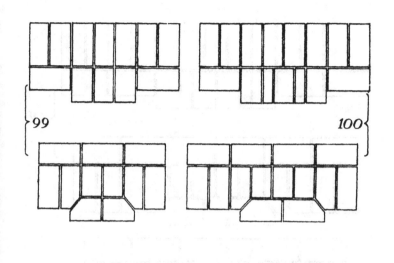

99 100

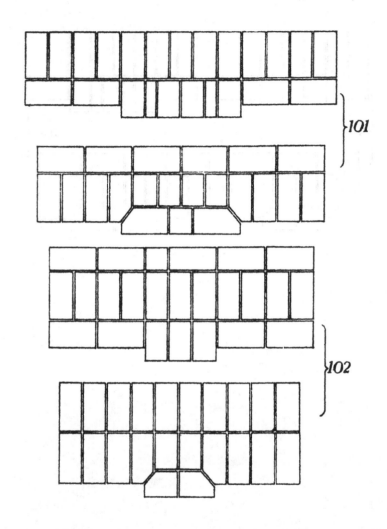

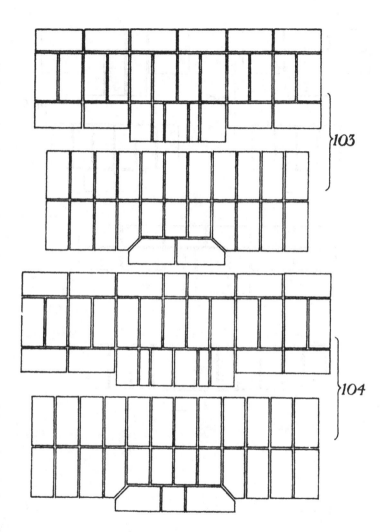

103

104

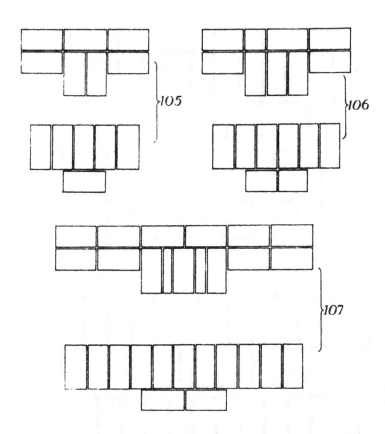

105

106

107

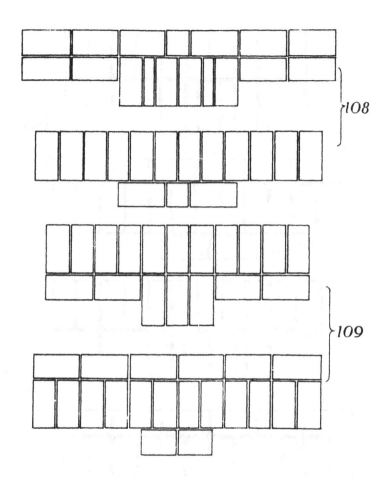

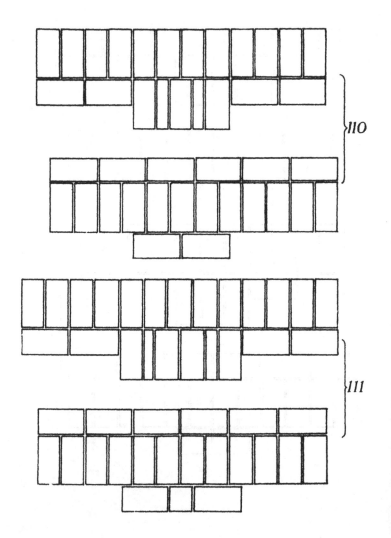

110

111

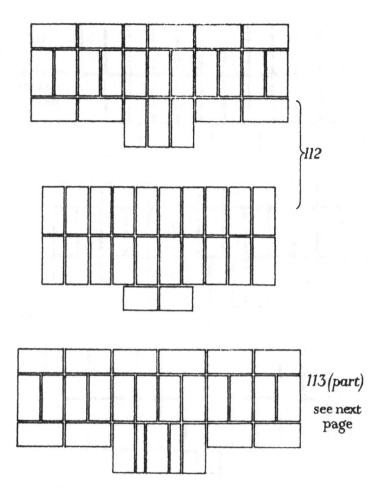

112

113 (part)

see next
page

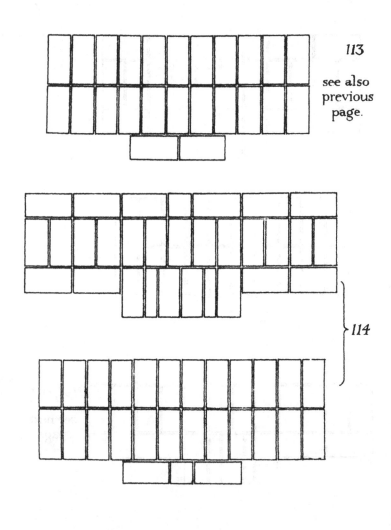

113

see also
previous
page.

114

ATTACHED PLAIN PIERS IN FLEMISH BOND WALLS

Figs. 115 to 148

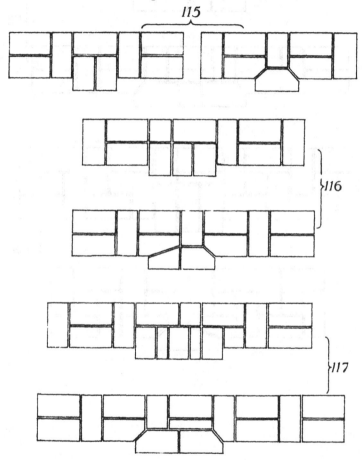

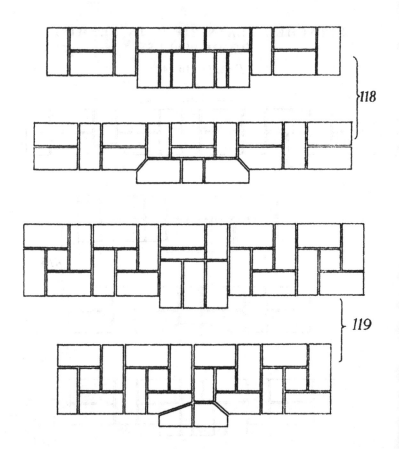

118

119

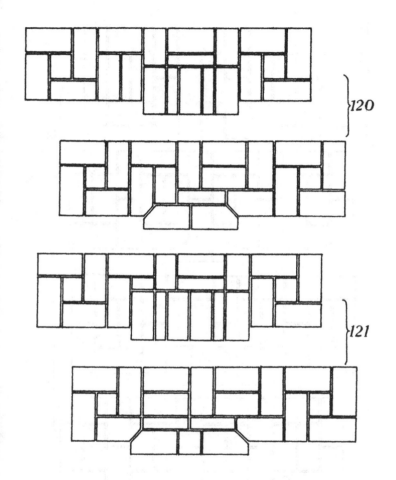

120

121

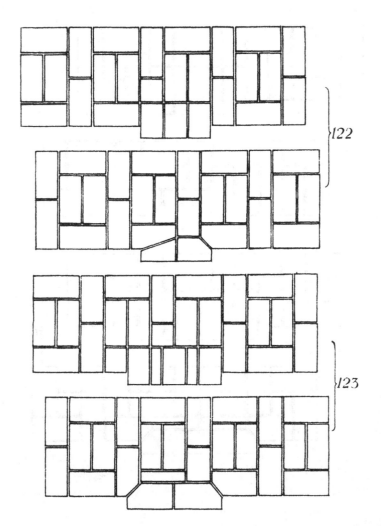

122

123

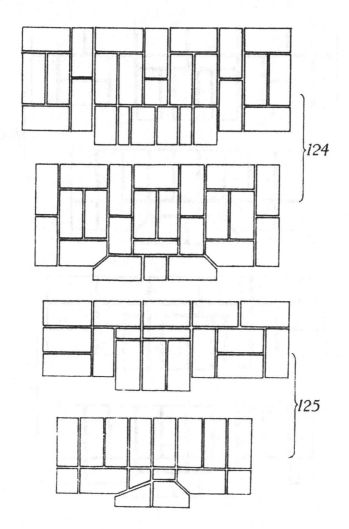

124

125

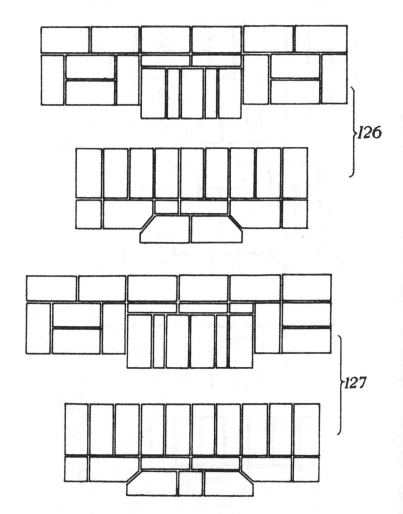

126

127

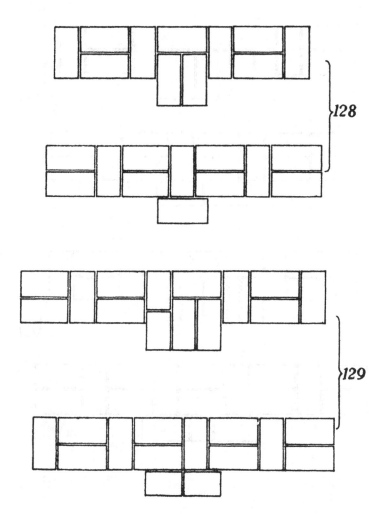

128

129

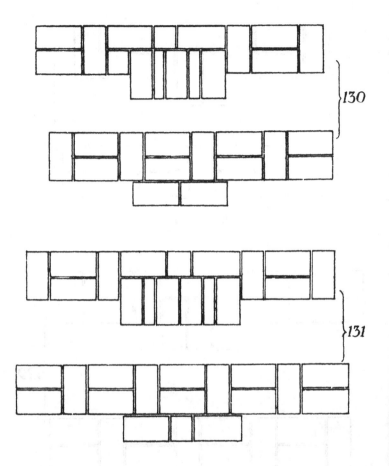

130

131

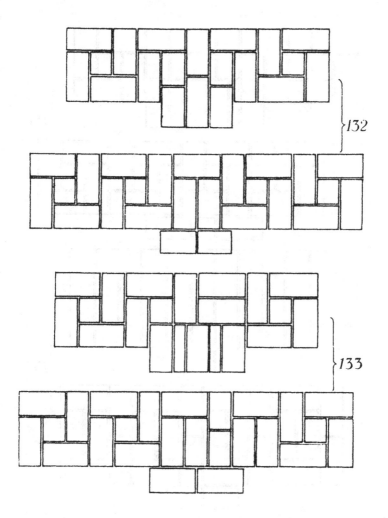

132

133

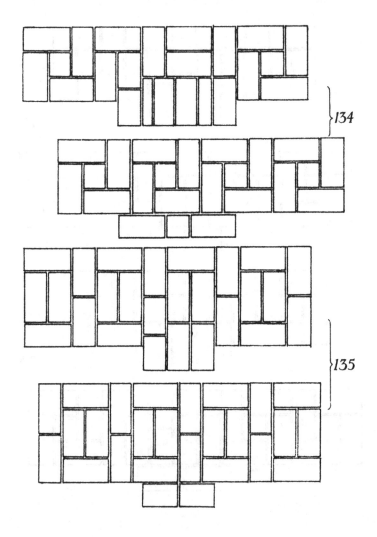

134

135

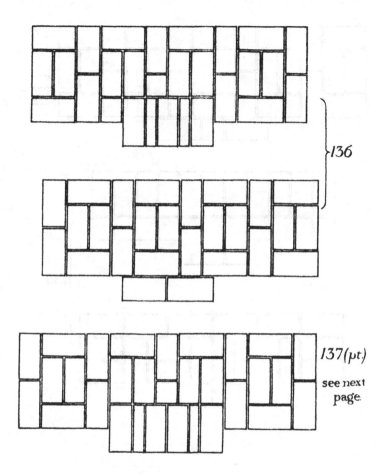

136

137 (pt.)

see next page

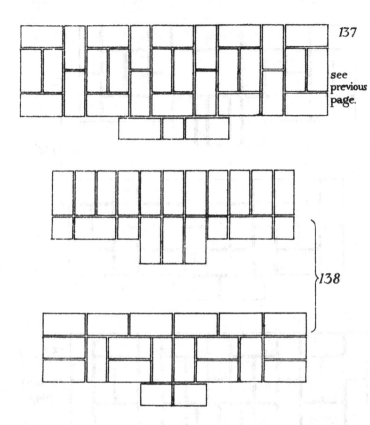

137

see
previous
page.

138

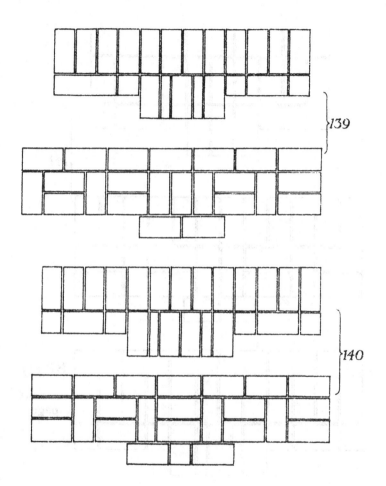

139

140

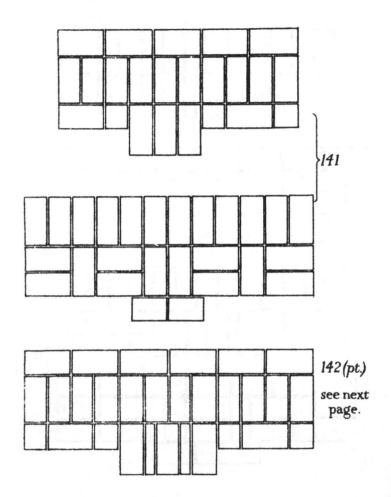

141

142 (pt.)

see next
page.

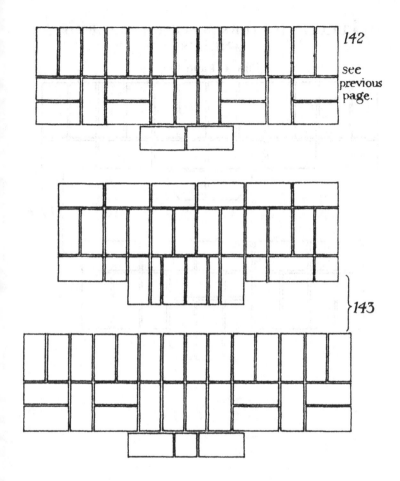

142

see previous page.

143

ATTACHED PIERS WITH CHAMFERED ANGLES

Figs. 144 and 145

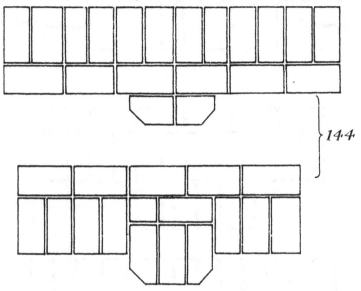

144

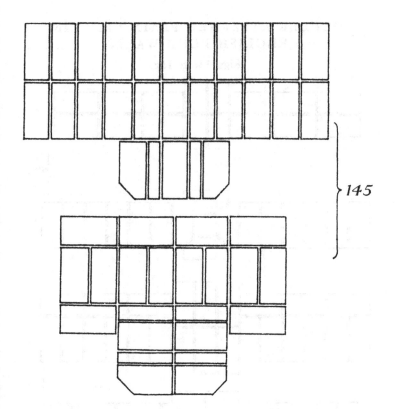

145

DOUBLE ATTACHED PLAIN PIERS IN ENGLISH BOND WALLS

Figs. 146 to 152

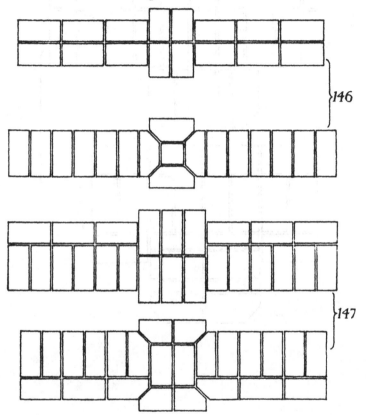

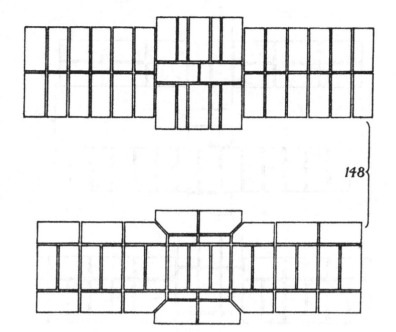

148

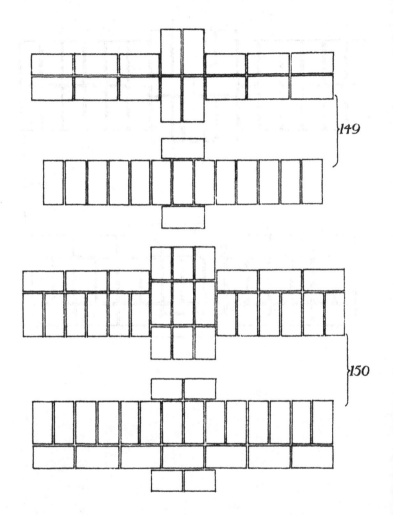

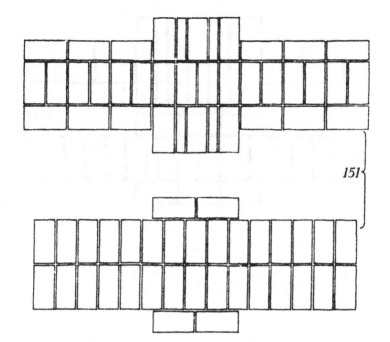

151

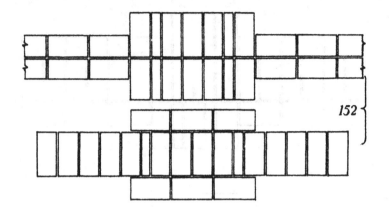

152

MISCELLANEOUS EXAMPLES OF
ATTACHED PIERS

Figs. 153 to 155

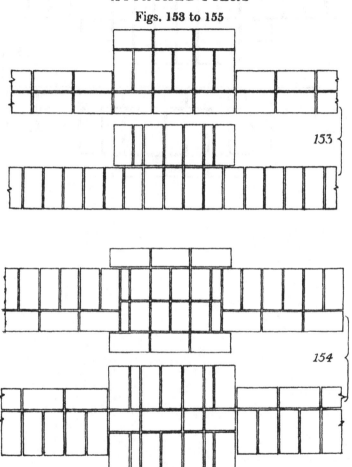

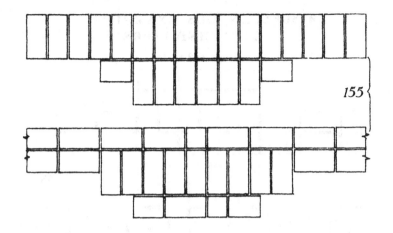

155

SQUARE JUNCTIONS IN ENGLISH BOND

Figs. 156 to 160

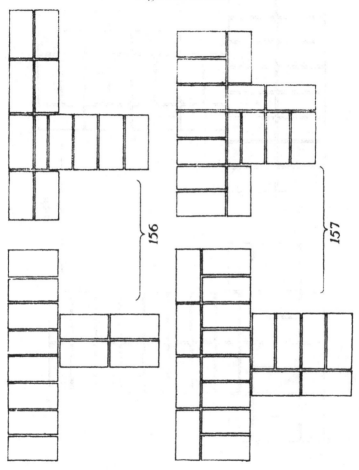

156

157

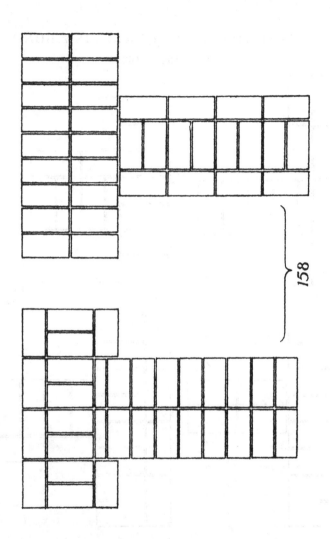

158

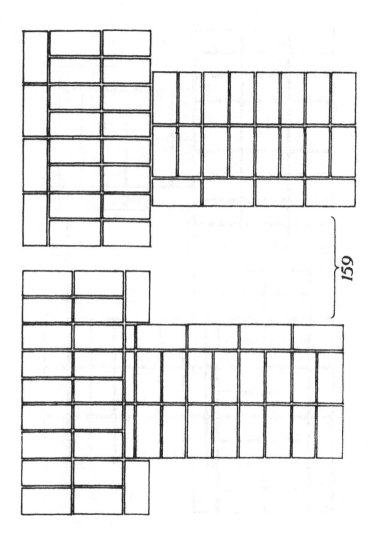

159

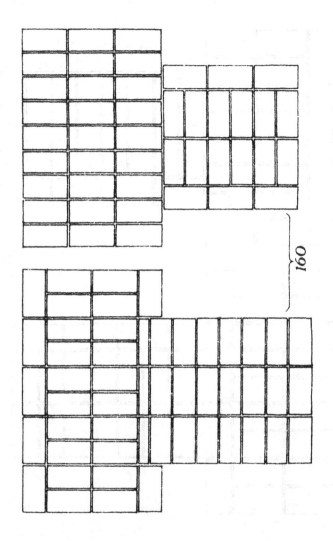

160

SQUARE JUNCTIONS IN FLEMISH BOND

Figs. 161 to 165

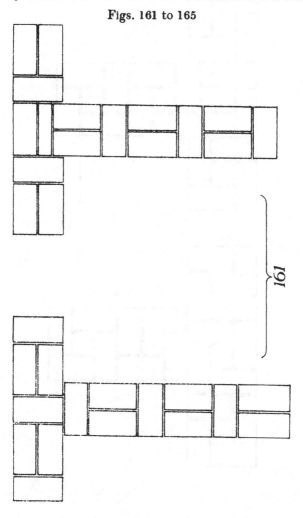

161

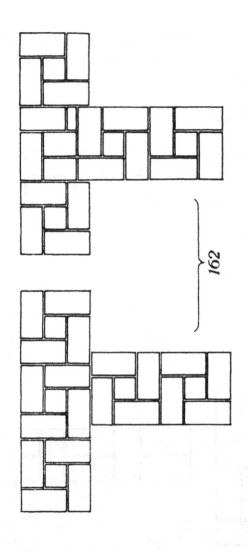

162

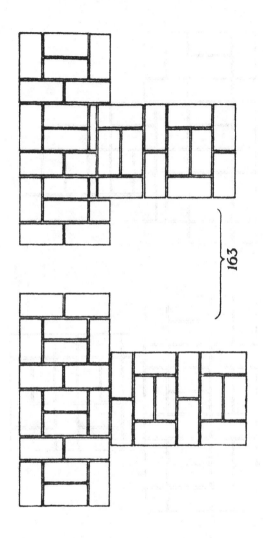

163

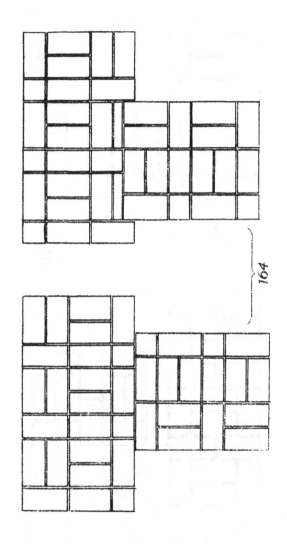

164

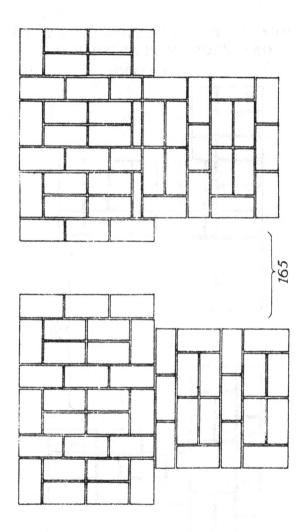

165

SQUARE JUNCTION TO WALLS WITH
ONE FACE IN FLEMISH BOND

Figs. 166 to 169

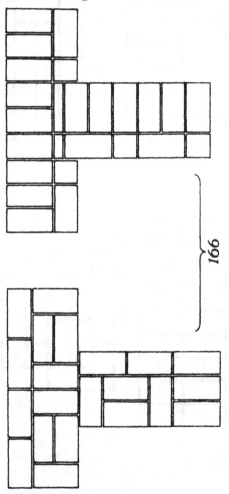

166

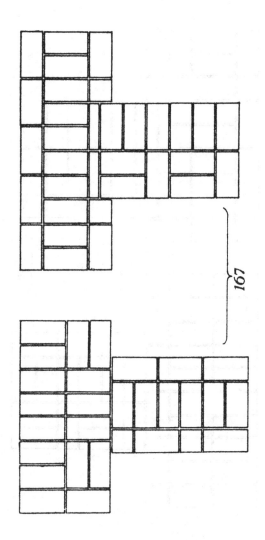

167

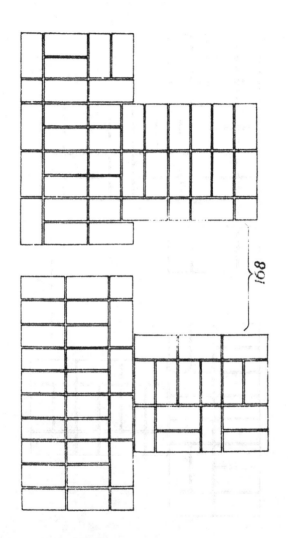

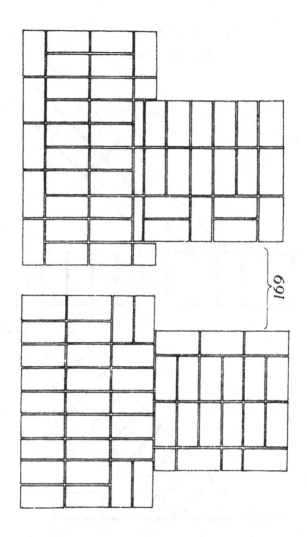

SPLAY JUNCTIONS
Figs. 170 to 175

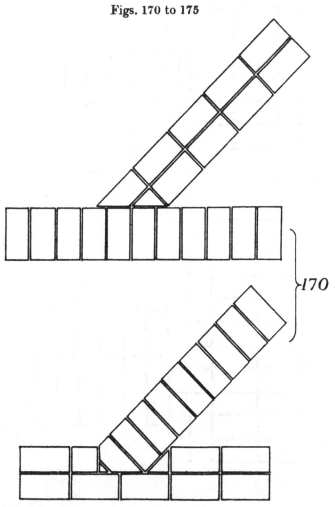

170

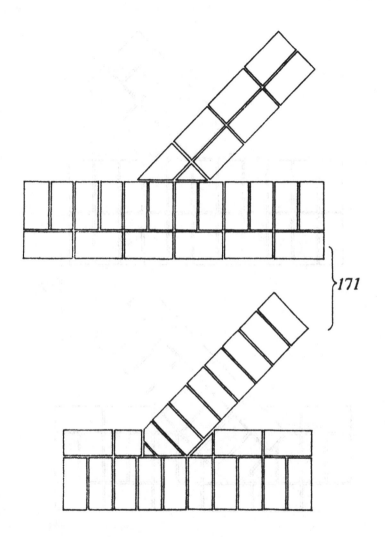

171

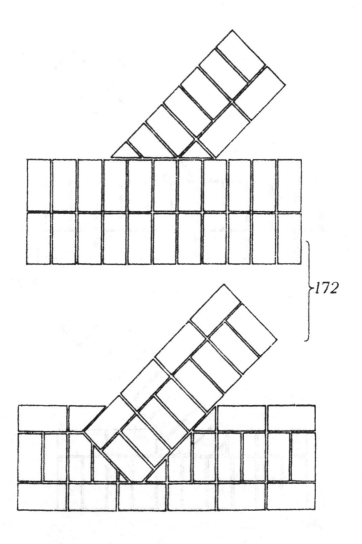

172

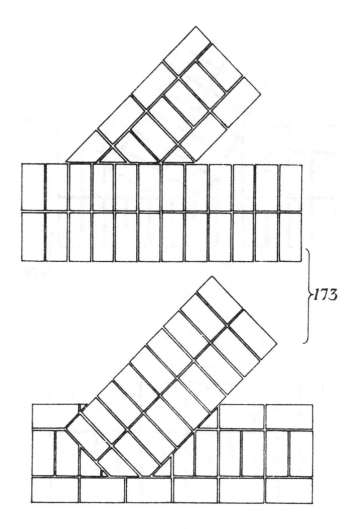

173

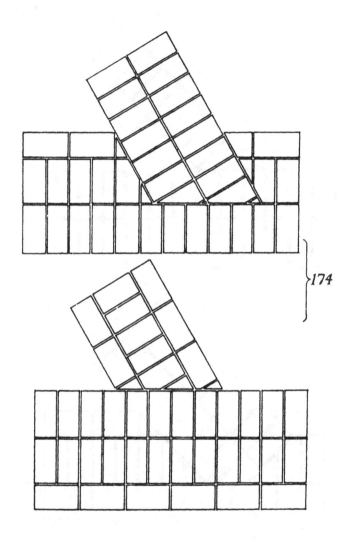

174

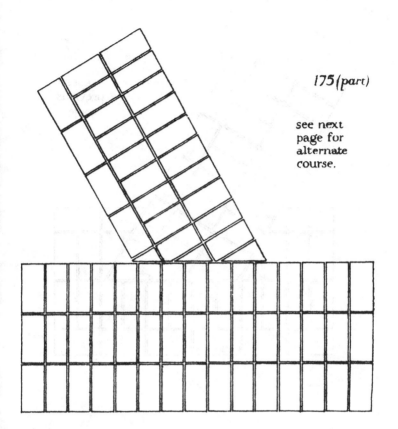

see next
page for
alternate
course.

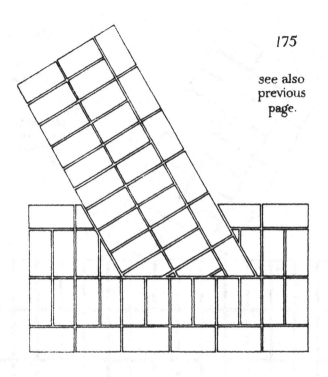

see also
previous
page.

CROSS JUNCTIONS
Figs. 176 to 181

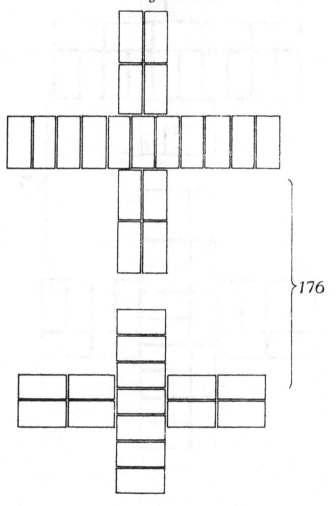

176

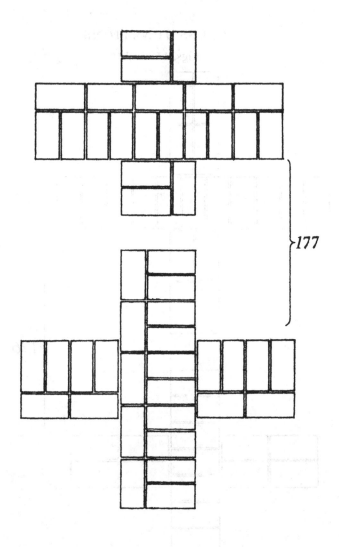

177

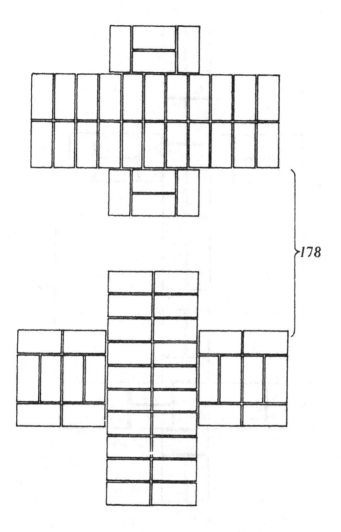

178

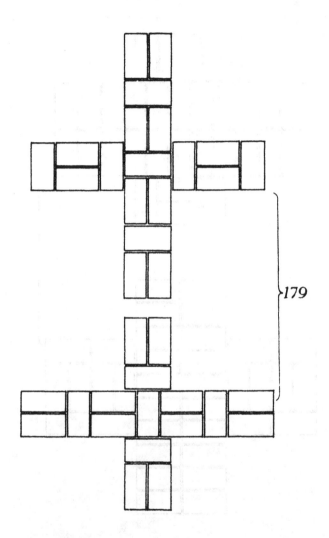

179

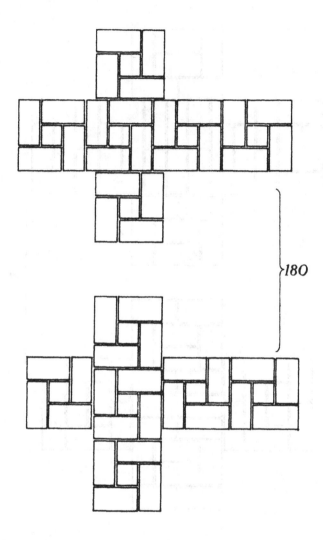

180

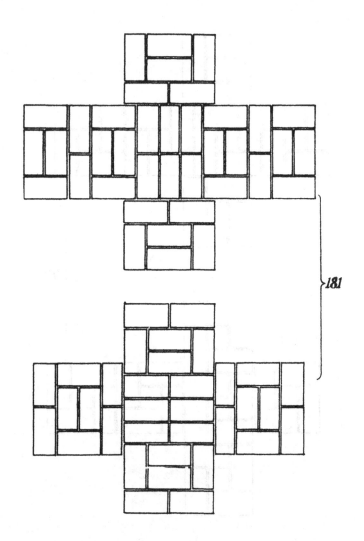

JUNCTIONS OF CURVED AND STRAIGHT WALLS

Figs. 182 to 185

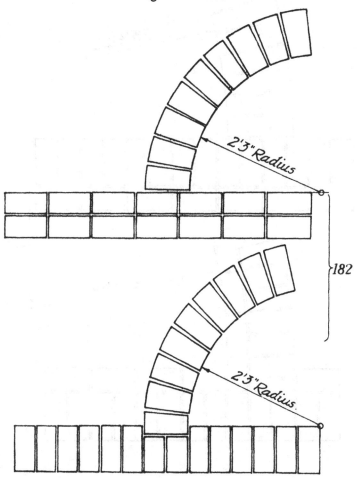

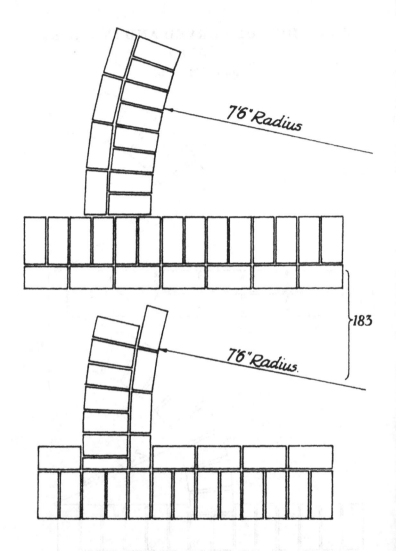

7'6" Radius

7'6" Radius

183

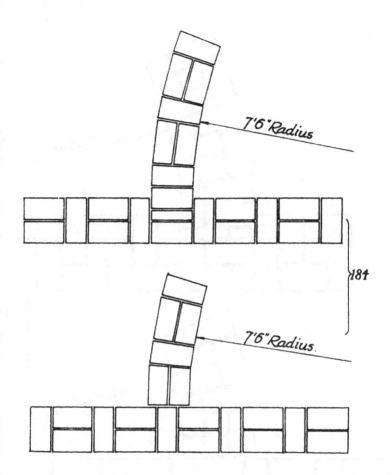

7'6" Radius

7'6" Radius.

184

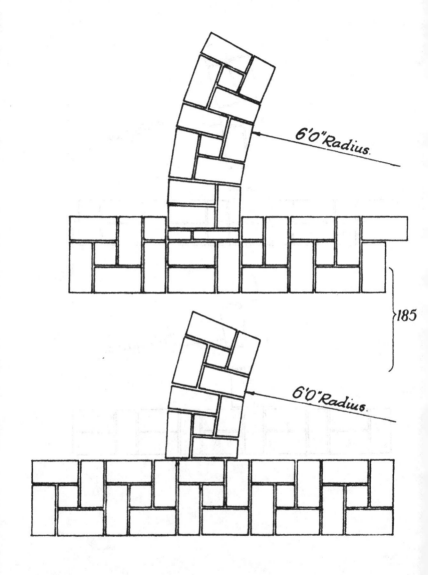

6'0" Radius.

185

6'0" Radius.

'BREAKS' OR 'RETURNS'

Figs. 186 to 194

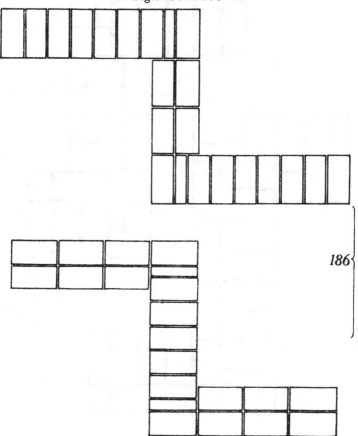

186

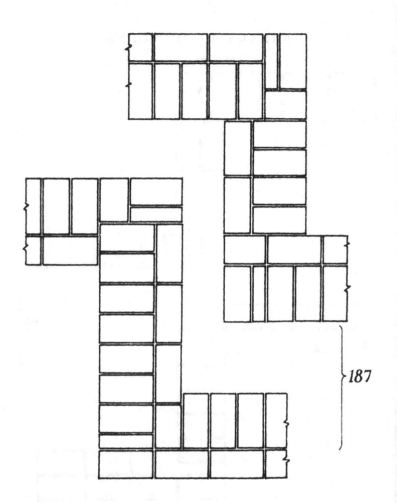

187

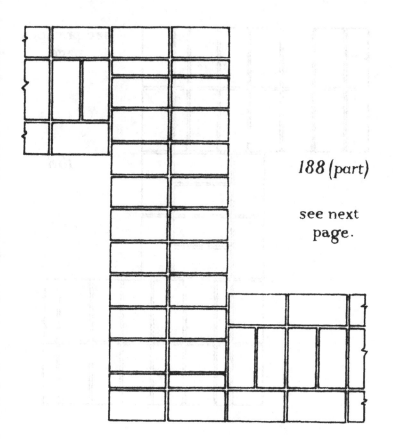

188 (part)

see next page.

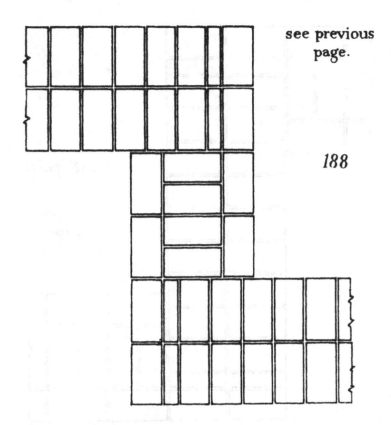

see previous
page.

188

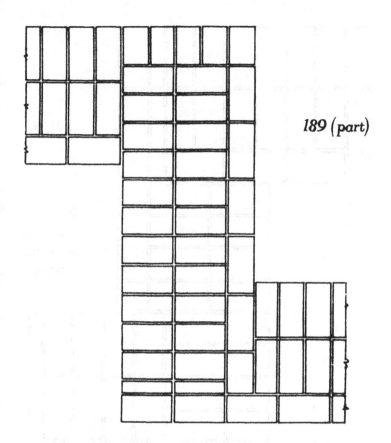

189 (part)

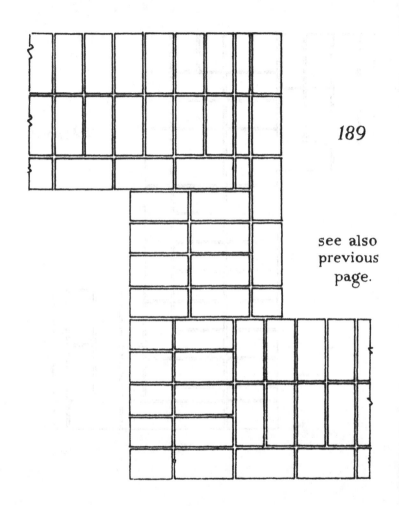

see also
previous
page.

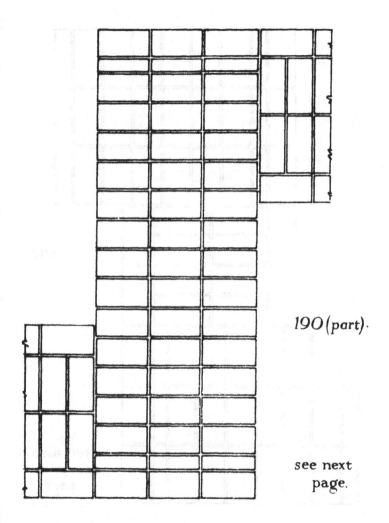

190 (part).

see next
page.

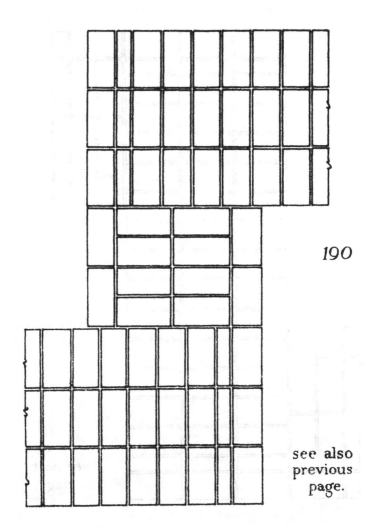

190

see also
previous
page.

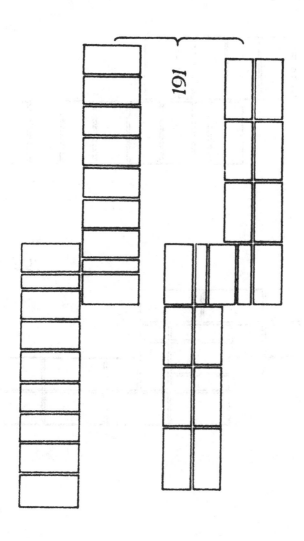

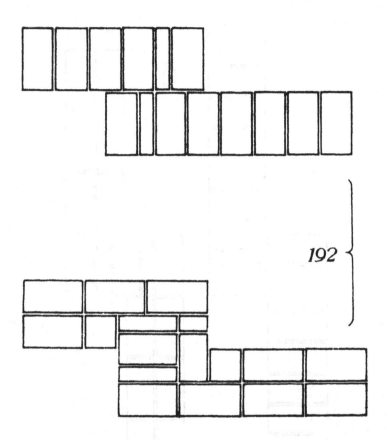

192

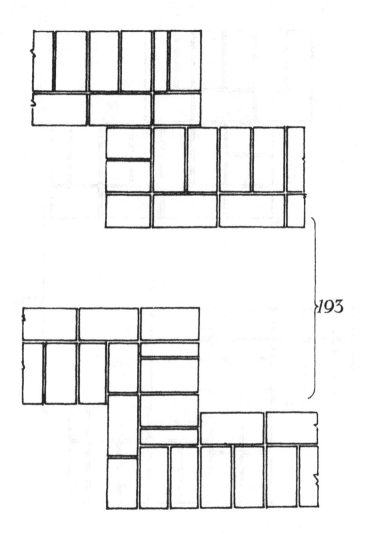

193

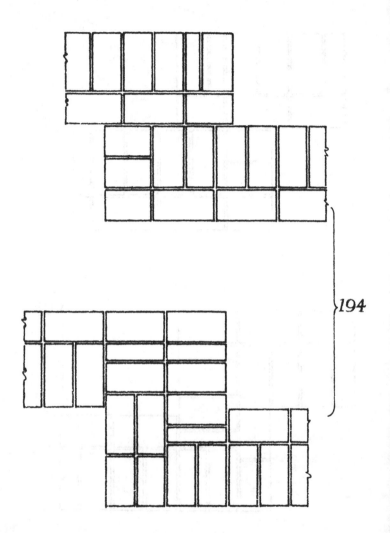

194

ISOLATED PIERS: ENGLISH BOND

Figs. 195 to 200

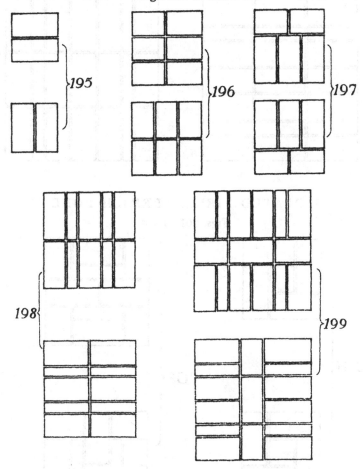

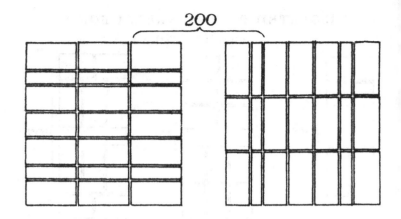

200

ISOLATED PIERS: FLEMISH BOND
Figs. 201 to 204

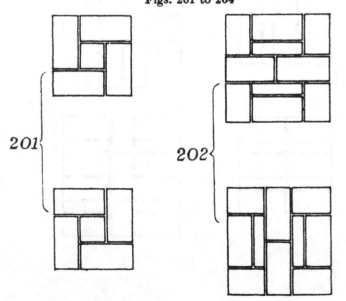

201

202

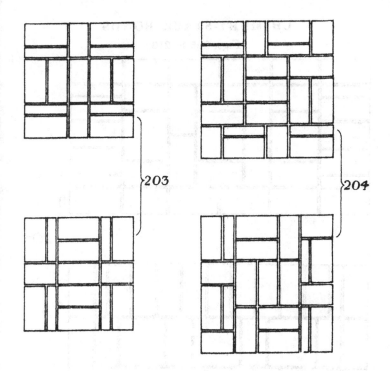

203

204

CHIMNEY-STACK BONDS

Figs. 205 to 210

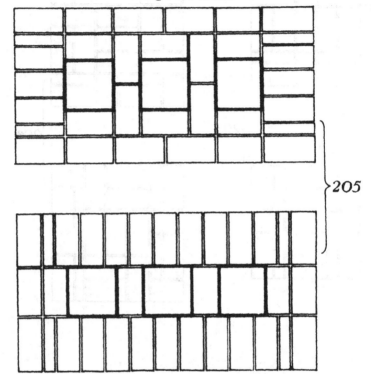

205

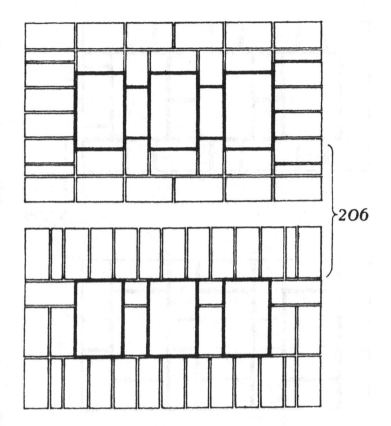

206

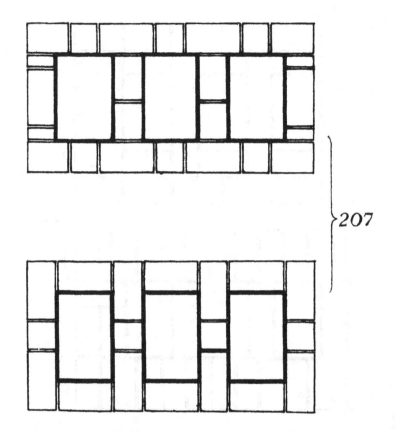

207

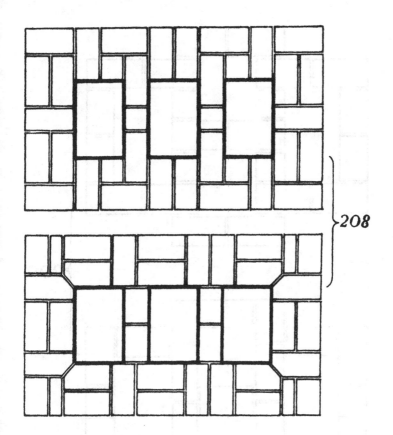

208

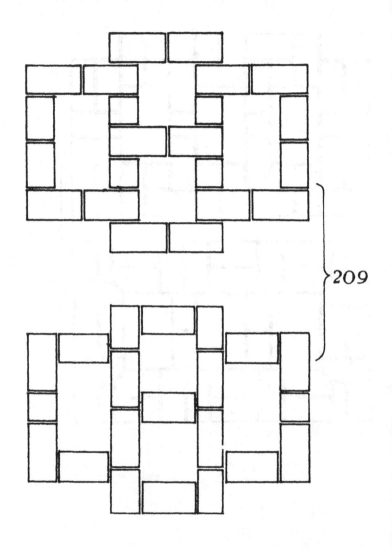

209

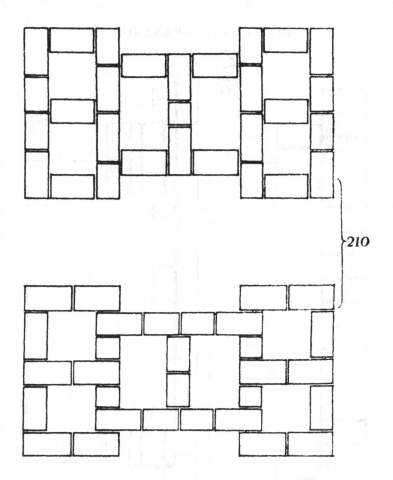

210

FIREPLACE OPENINGS
Figs. 211 to 213

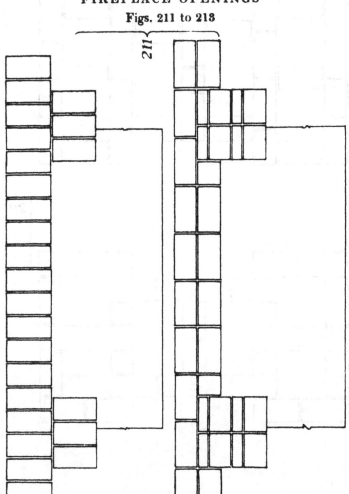

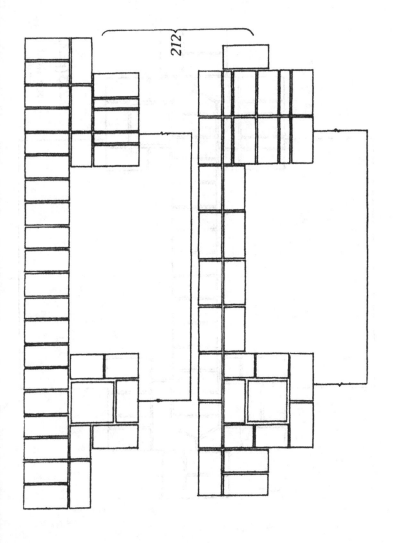

212

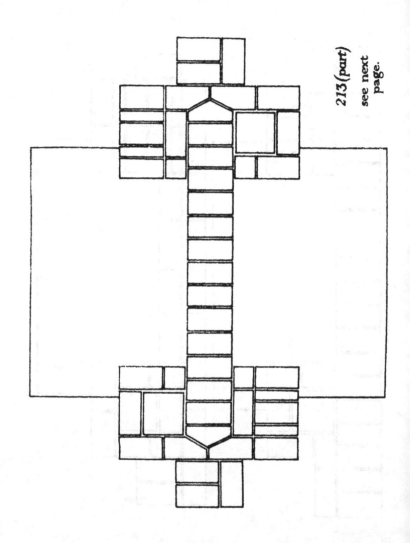

213 (part)
see next page.

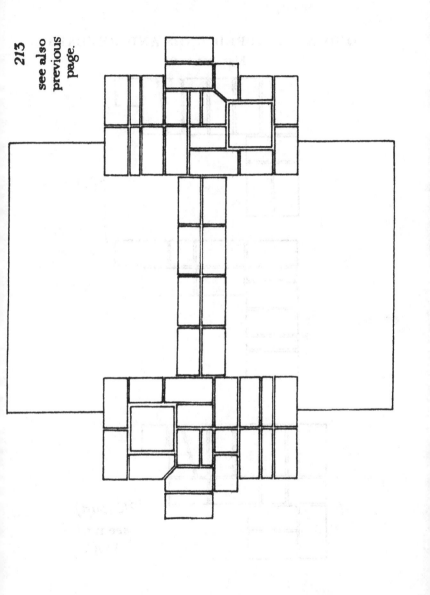

QUOINS, STOPPED ENDS AND REVEALS
Figs. 214 to 217

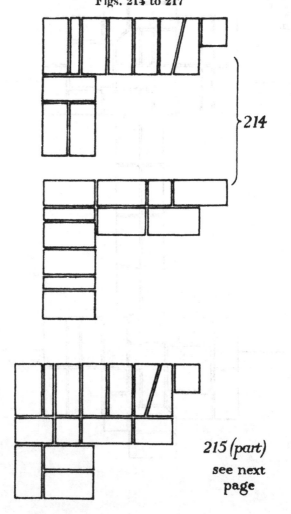

214

215 (part)
see next
page

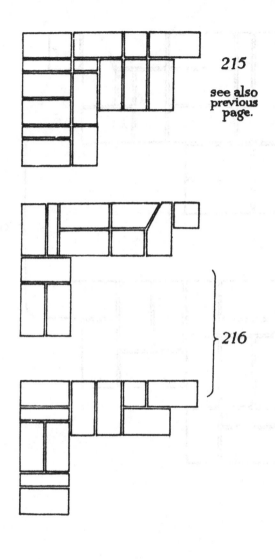

215

see also
previous
page.

216

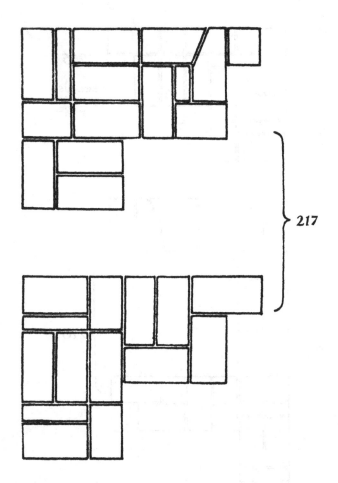

217

STEPPED REVEALS
Figs. 218 to 225

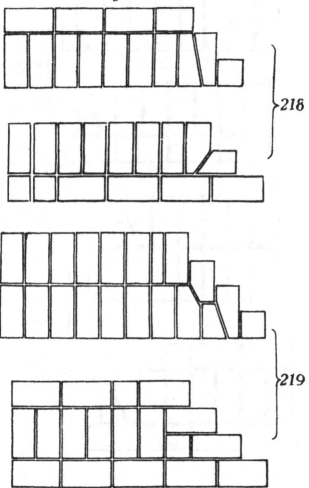

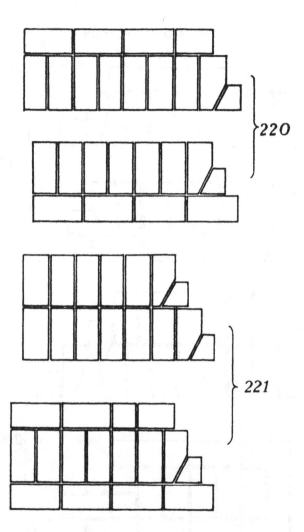

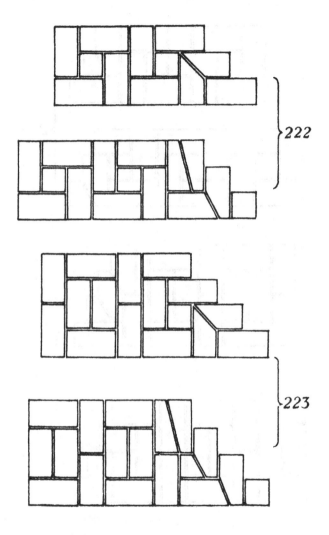

222

223

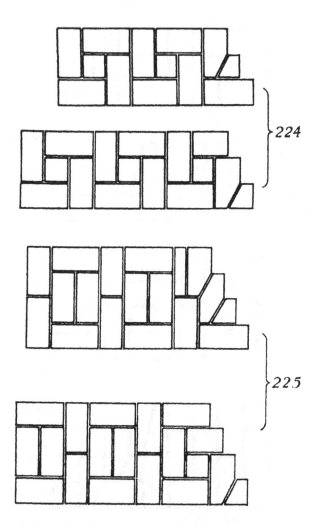

224

225

PROJECTING QUOINS
Figs. 226 to 230

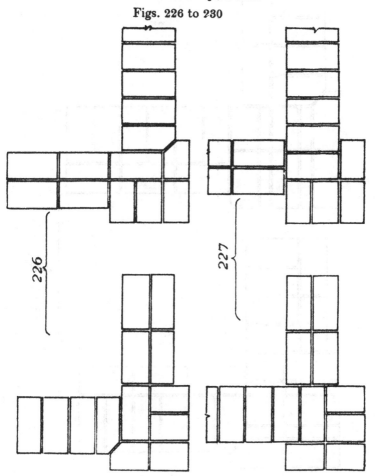

226

227

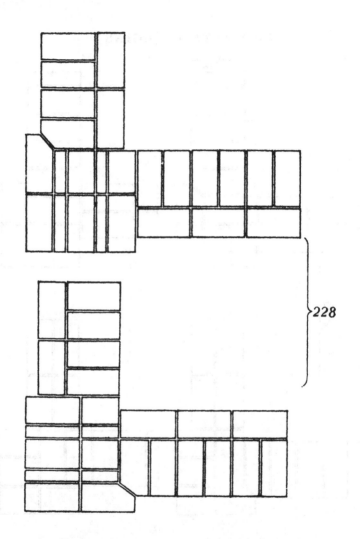

228

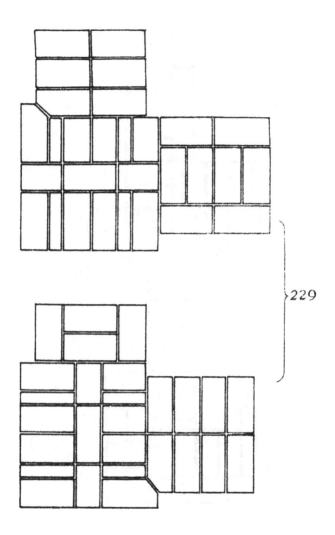

229

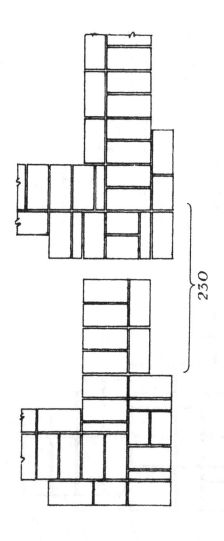

230

MISCELLANEOUS EXAMPLES OF
BONDING
Figs. 231 to 258

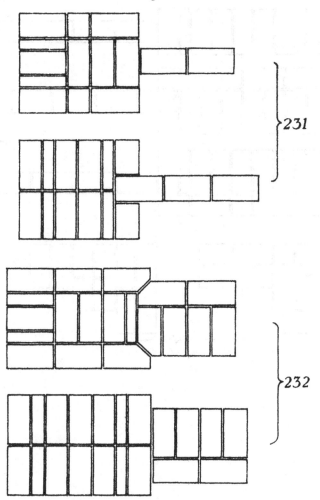

231

232

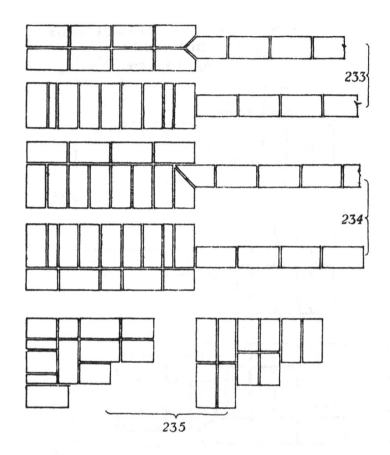

233

234

235

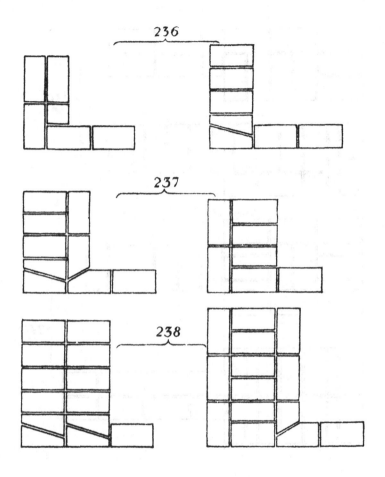

236

237

238

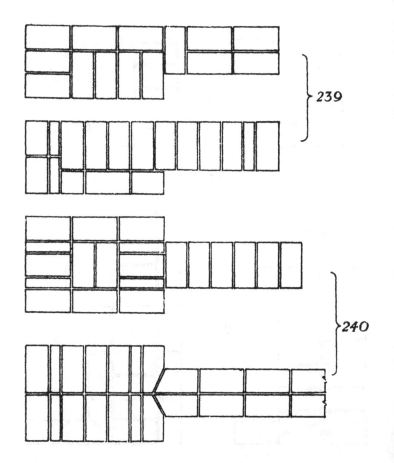

239

240

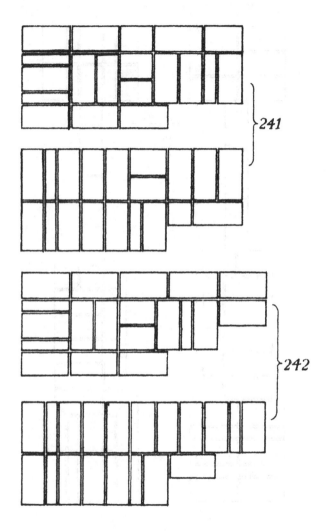

241

242

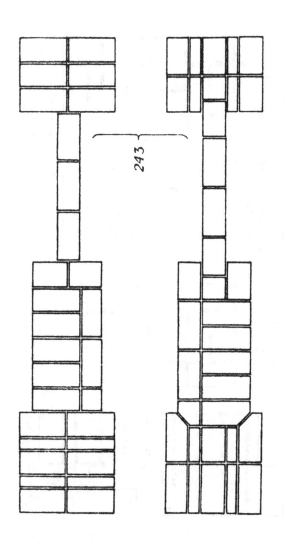

243

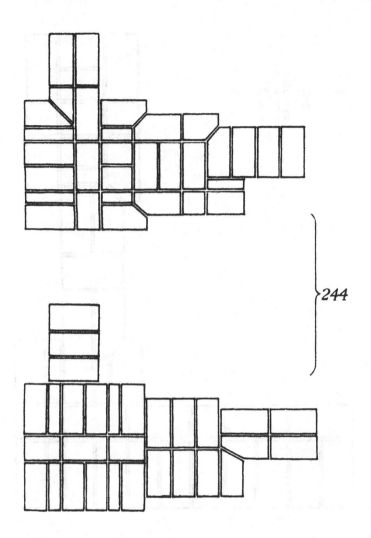

244

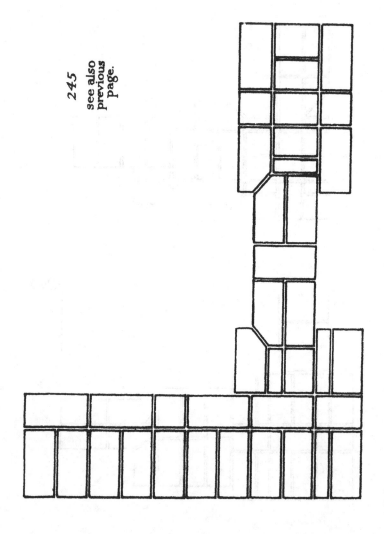

245
see also previous page.

245 (part)
see next
page

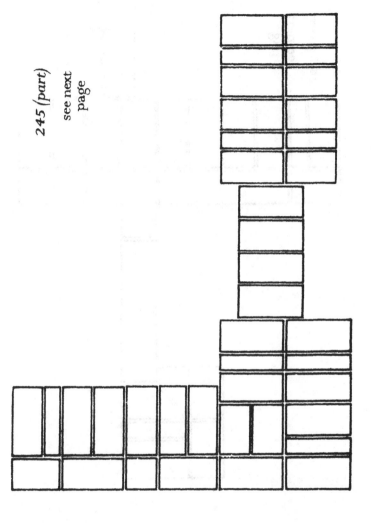

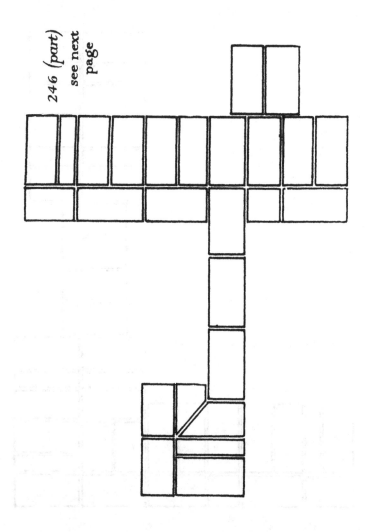

246 (part)
see next page

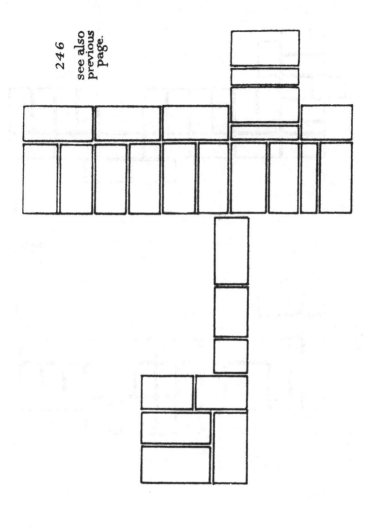

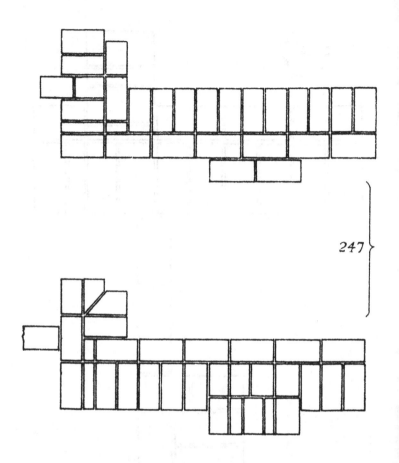

247

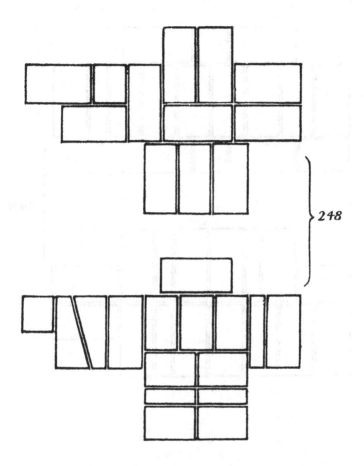

248

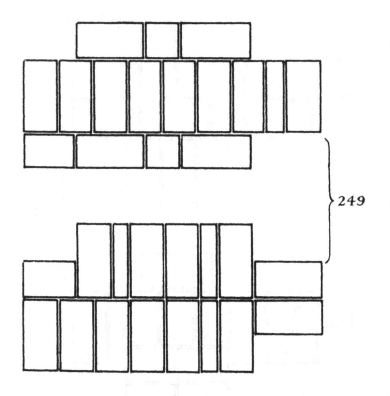

249

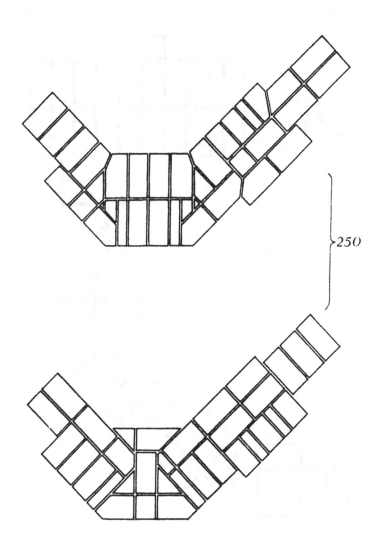

250

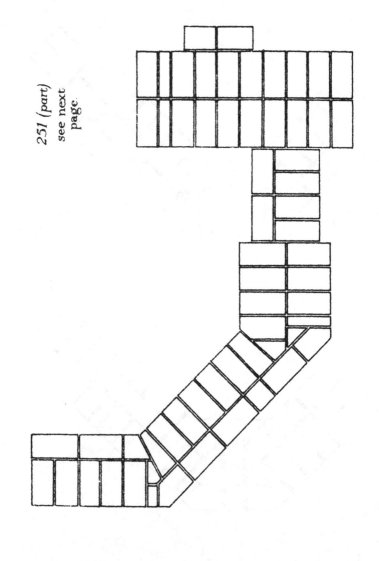

251 (part)
see next
page.

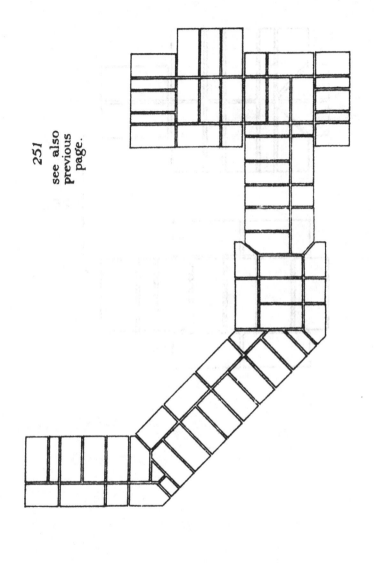

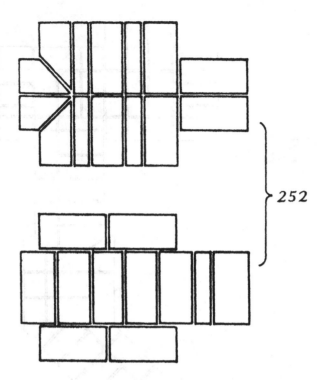

252

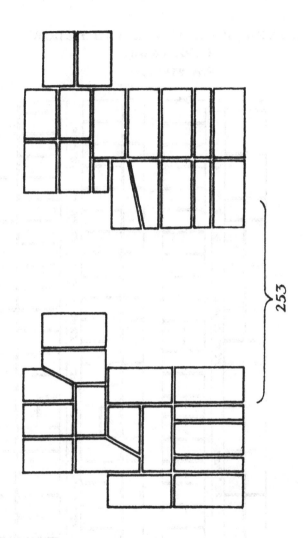

253

BROKEN BOND ABOVE AND BELOW
OPENINGS
Figs. 254 to 260

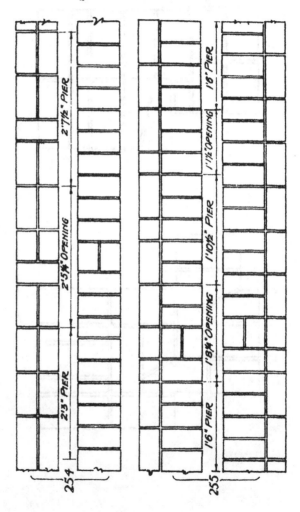

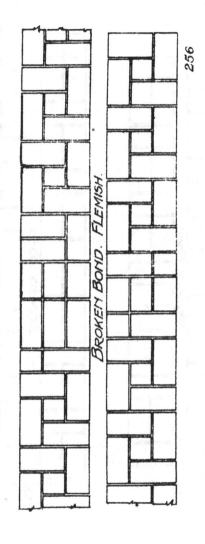

BROKEN BOND. FLEMISH.

256

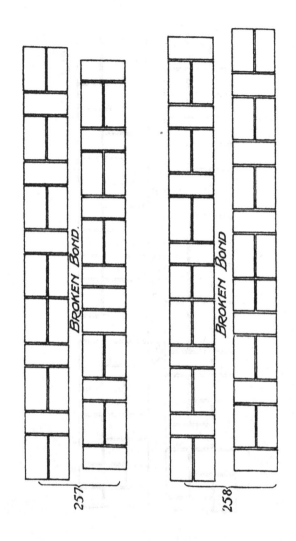

BROKEN BOND.

257

BROKEN BOND

258

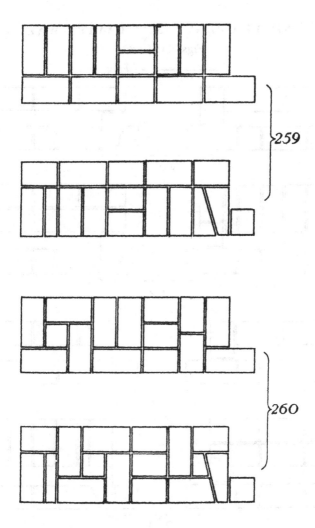

259

260

BOND AT OPENINGS IN CAVITY WALLS

Figs. 261 to 264

⊢————— = 1 Foot

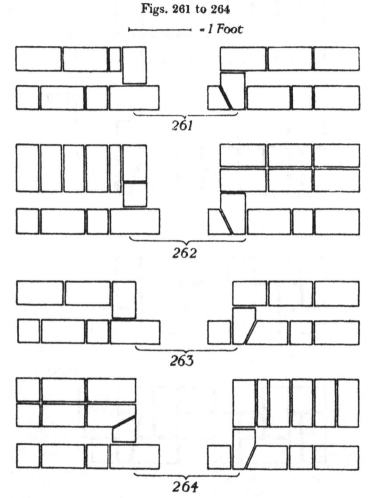

261

262

263

264

SQUARE-ANGLED BAY WINDOWS
Figs. 265 and 266

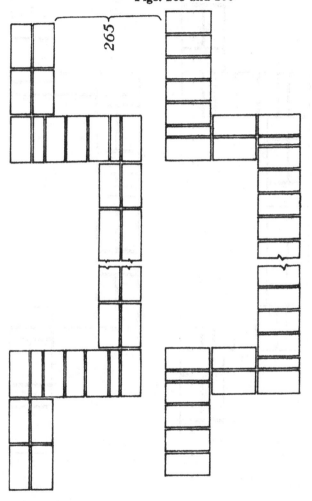

266

FOOTINGS TO QUOINS OF VARIOUS WALLS

Figs. 267 to 269

267

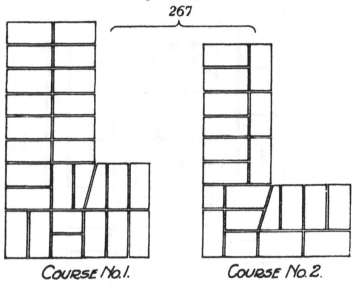

Course No. 1.   *Course No. 2.*

FOOTINGS TO QUOIN OF 9" WALL.

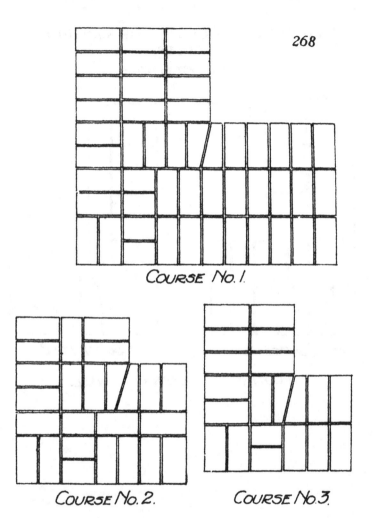

268

COURSE No. 1.

COURSE No. 2.

COURSE No. 3.

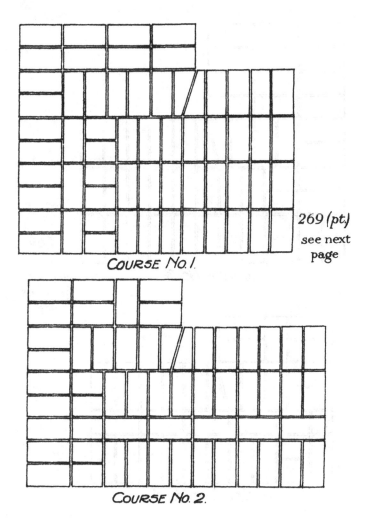

COURSE No. 1.

269 (pt.)
see next
page

COURSE No. 2.

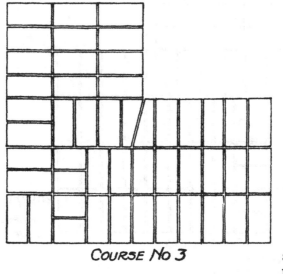

COURSE No 3

see also
previous
page.

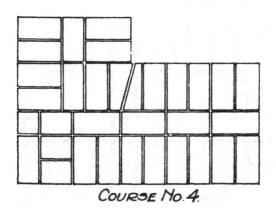

COURSE No. 4.

270

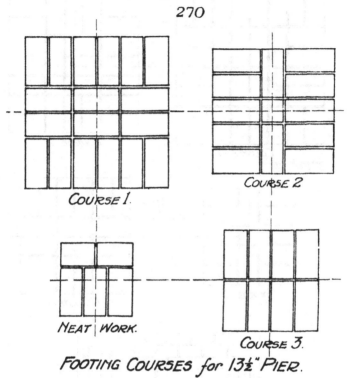

COURSE 1.

COURSE 2

NEAT WORK.

COURSE 3.

FOOTING COURSES for 13½" PIER.

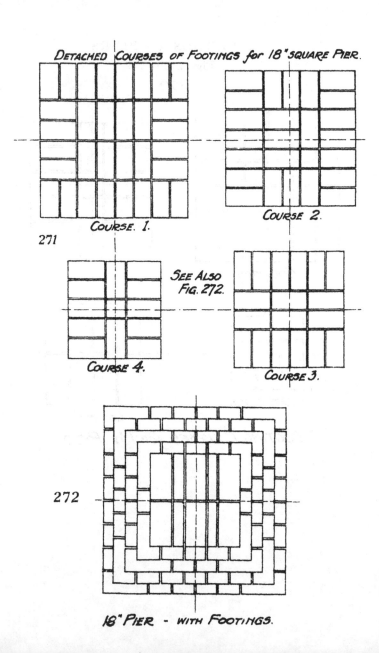

DETACHED COURSES OF FOOTINGS for 18" SQUARE PIER.

COURSE. 1.

COURSE 2.

271

SEE ALSO FIG. 272.

COURSE 4.

COURSE 3.

272

18" PIER - WITH FOOTINGS.

FOOTINGS TO OBTUSE AND ACUTE QUOINS

Figs. 273 and 274

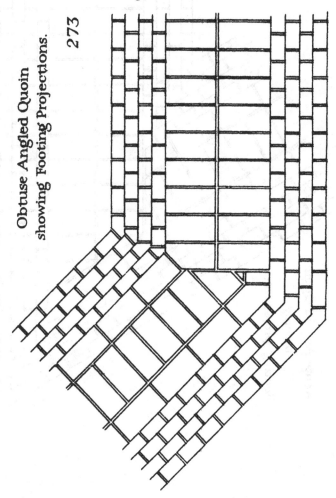

Obtuse Angled Quoin showing Footing Projections.

273

Acute Angled Quoin
showing
Footing projections

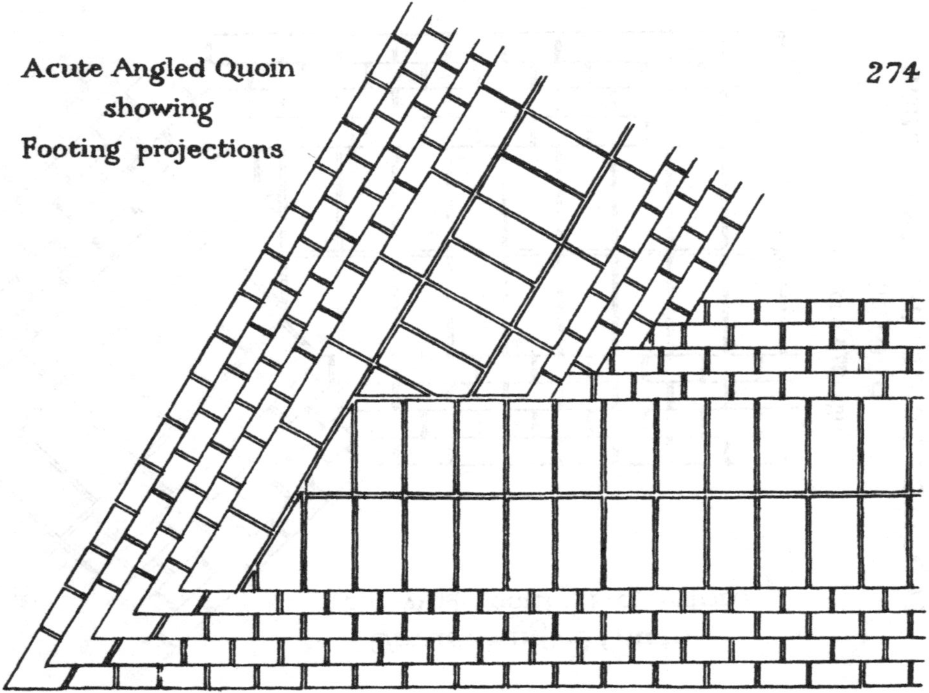

274

SPECIAL BONDING FOR VARIOUS EXAMPLES

Figs. 275 to 278

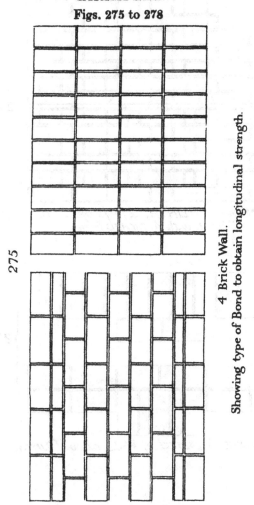

275

4 Brick Wall.

Showing type of Bond to obtain longitudinal strength.

ELEVATION.

276 (a)

CHAMFERED PLINTH, 2¼" projection,
in 9" WALL with ATTACHED PIER.

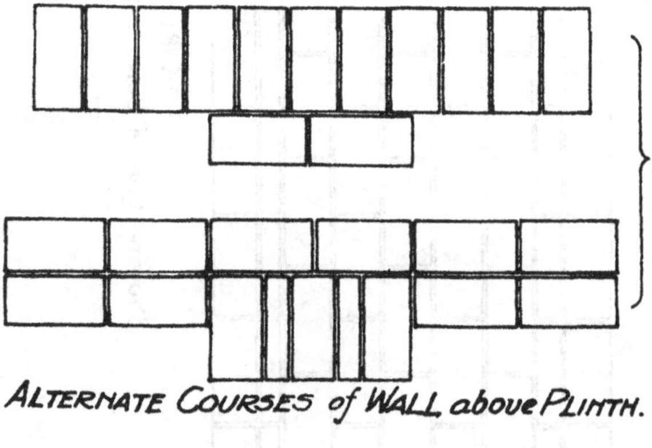

276 (b)

ALTERNATE COURSES of WALL above PLINTH.

SEE NEXT PAGE FOR PLINTH COURSES

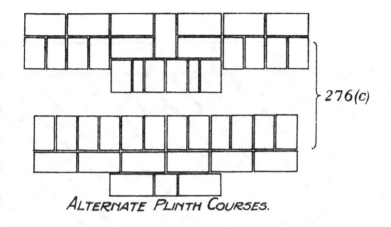

276(c)

ALTERNATE PLINTH COURSES.

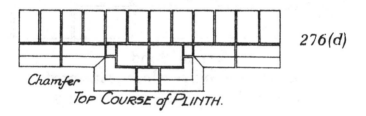

276(d)

Chamfer

TOP COURSE OF PLINTH.

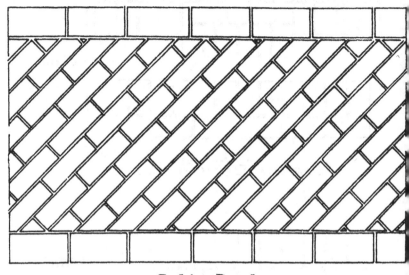

Raking Bond.
Plan of *one course* which is inserted
at intervals in thick walls – or in footing courses.

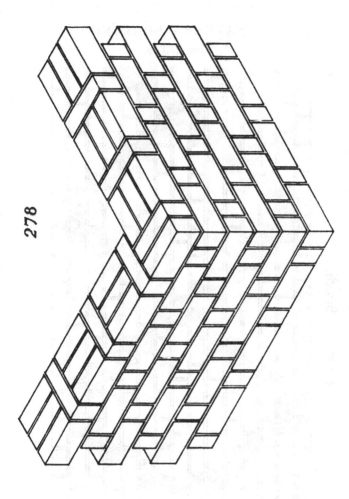

Rat Trap Bonding in Right Angled 9" Quoin.

PATTERN BONDING
Figs. 279 to 280

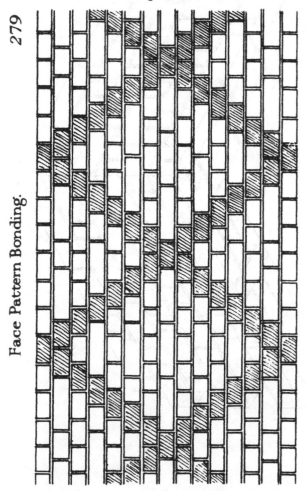

279

Face Pattern Bonding.

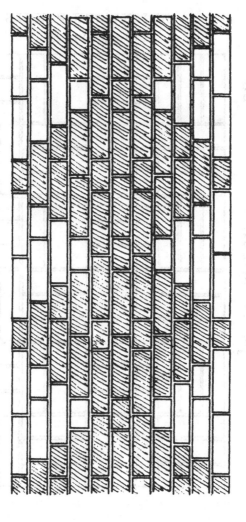

Face Pattern Bonding.

280

281

Diaper-bond Panel.

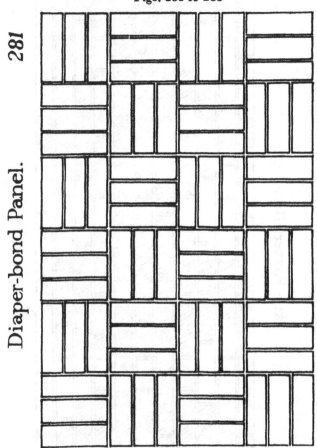

Brick Panel formed with "headers".

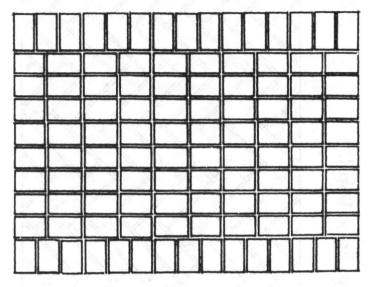

282

Panel in Herring-bone Bond.

283

EXAMPLES FOR PRACTICE

See next page for plinth courses

Stretching Bond

EXAMPLES FOR PRACTICE

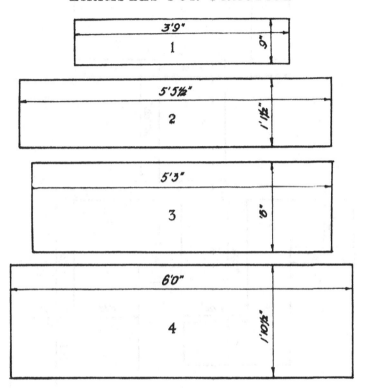

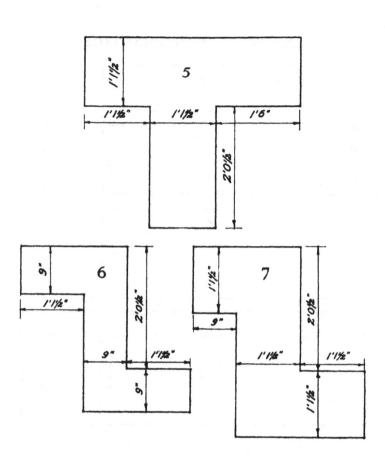

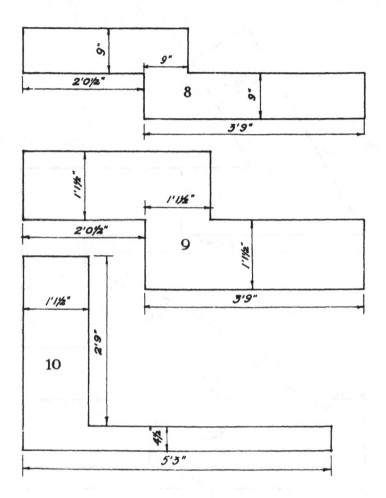

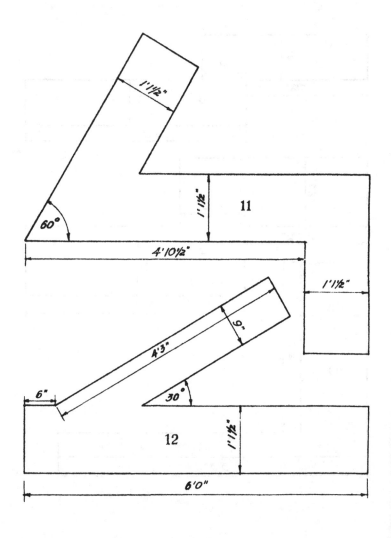

11

1'1½"

1'1½"

60°

4'10½"

1'1½"

12

6"

4'3"

6"

30°

1'1½"

6'0"

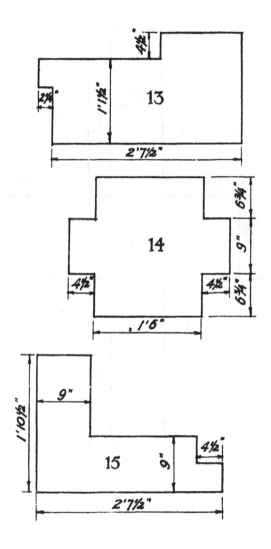

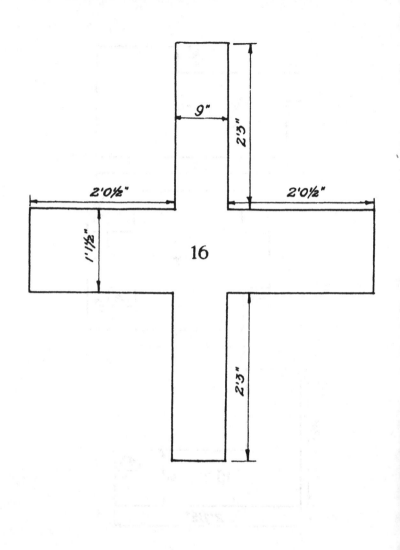

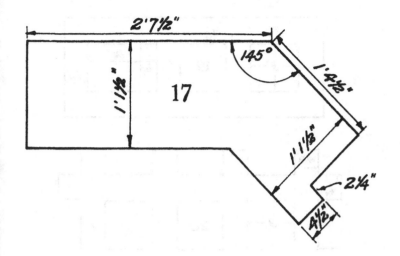

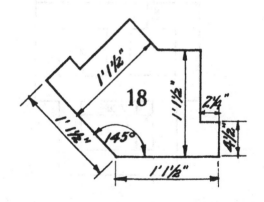

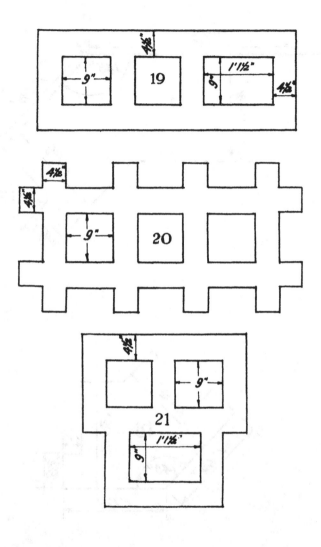

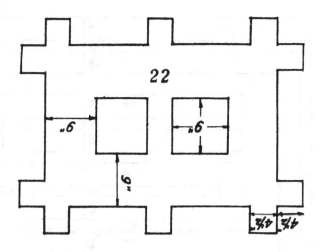

22

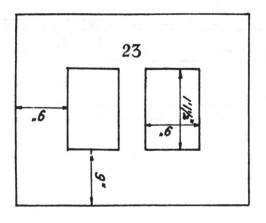

23

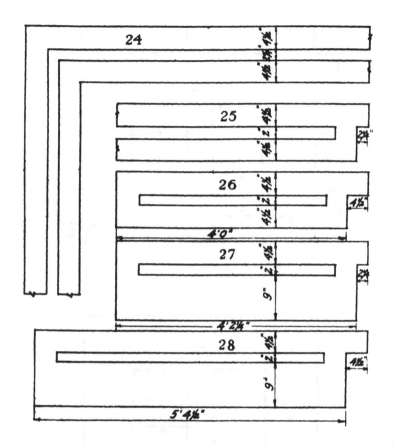

24

1¼"
6½"
4½"

25

4½" 2' 4½"

2¼"

26

4½"
4½" 2' 4½"

4½"

4'0"

27

4½"
2' 4½"

9"

2¼"

4'2¼"

28

4½"
2' 4½"

9"

4½"

5'4½"

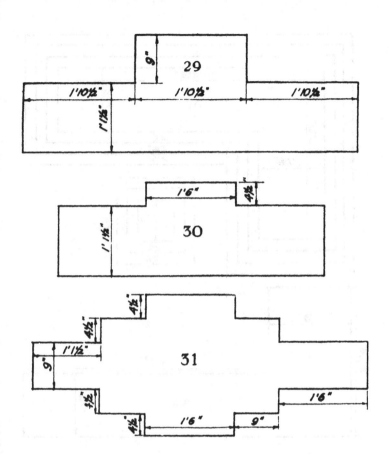

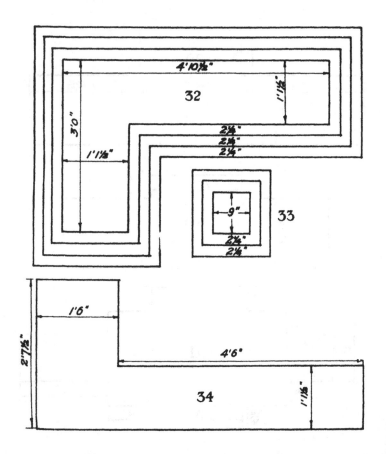

32

4'10½"

1'1½"

3'0"

1'1½"

2½"
2½"
2½"

33

9"

2½"
2½"

34

1'6"

2'7½"

4'6"

1'1½"

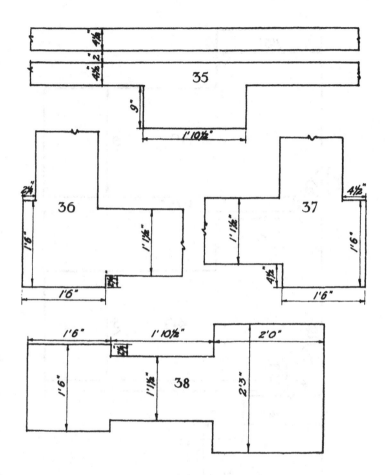

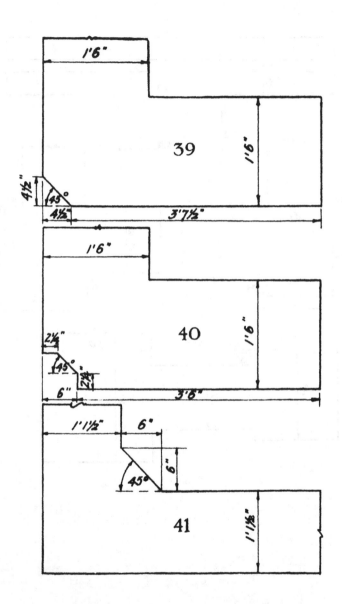

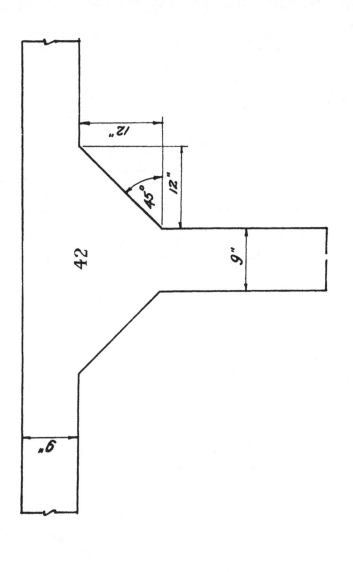

42

12"

12"

45°

9"

9"

Printed in the United States
By Bookmasters